**Springer Theses**

Recognizing Outstanding Ph.D. Research

## Aims and Scope

The series "Springer Theses" brings together a selection of the very best Ph.D. theses from around the world and across the physical sciences. Nominated and endorsed by two recognized specialists, each published volume has been selected for its scientific excellence and the high impact of its contents for the pertinent field of research. For greater accessibility to non-specialists, the published versions include an extended introduction, as well as a foreword by the student's supervisor explaining the special relevance of the work for the field. As a whole, the series will provide a valuable resource both for newcomers to the research fields described, and for other scientists seeking detailed background information on special questions. Finally, it provides an accredited documentation of the valuable contributions made by today's younger generation of scientists.

## Theses are accepted into the series by invited nomination only and must fulfill all of the following criteria

- They must be written in good English.
- The topic should fall within the confines of Chemistry, Physics, Earth Sciences, Engineering and related interdisciplinary fields such as Materials, Nanoscience, Chemical Engineering, Complex Systems and Biophysics.
- The work reported in the thesis must represent a significant scientific advance.
- If the thesis includes previously published material, permission to reproduce this must be gained from the respective copyright holder.
- They must have been examined and passed during the 12 months prior to nomination.
- Each thesis should include a foreword by the supervisor outlining the significance of its content.
- The theses should have a clearly defined structure including an introduction accessible to scientists not expert in that particular field.

More information about this series at http://www.springer.com/series/8790

Biao Li

# Studies on Anionic Redox in Li-Rich Cathode Materials of Li-Ion Batteries

Doctoral Thesis accepted by
the Peking University, Beijing, P. R. China

Springer

*Author*
Dr. Biao Li
Mechanics (Energy and Resources Engineering), College of Engineering
Peking University
Beijing, P. R. China

*Supervisor*
Prof. Dingguo Xia
Beijing Key Laboratory of Theory and Technology for Advanced Batteries Materials, College of Engineering
Peking University
Beijing, P. R. China

ISSN 2190-5053 ISSN 2190-5061 (electronic)
Springer Theses
ISBN 978-981-13-2846-6 ISBN 978-981-13-2847-3 (eBook)
https://doi.org/10.1007/978-981-13-2847-3

Library of Congress Control Number: 2018957250



This Springer imprint is published by the registered company Springer Nature Singapore Pte Ltd.
The registered company address is: 152 Beach Road, #21-01/04 Gateway East, Singapore 189721, Singapore

# Supervisor's Foreword

Nowadays, the research on cathode materials of lithium-ion batteries is focusing on exploring high-specific-capacity electrodes. Lithium-rich layered oxides can exhibit abnormal capacities due to the joint participation of cation and anion in the redox process. It is of great importance to study and understand the intrinsic anionic redox mechanism in purpose to design and modify new kind of high-capacity electrodes. Here, based on the deep understanding of anionic redox, we have tuned the redox process of lithium-rich electrode for achieving better electrochemical performance and uncovered the stability mechanism of the local structure upon anionic redox in lithium-rich cathode materials, as well as the activation mechanism of anionic redox in these electrode systems. The main contents and conclusions include:

1. We have fabricated a kind of polyanion-modified lithium-rich manganese-based layered oxide, by introducing boracic polyanions into the layered structure to tune its electronic structure. The as-prepared electrode exhibits excellent cycling performance, specifically, showing 300 mAh $g^{-1}$ when cycled to 80th cycle under 0.1C rate current and 89% retention cycled to 300th cycle under 0.3C rate, with also higher thermal stability and charge/discharge potentials. Combined series of techniques, we have shown that the improvement should be ascribed to the incorporation of boracic polyanions, which lowers the overlap of M 3d (M = transition metal) and O 2p states, thus reducing the oxygen oxidation and resulting in a more stable structure.
2. Based on the model of $Li_2RuO_3$, the geometric and electronic structural evolutions of the lithium-rich high-capacity electrode were systematically investigated, with the aid of in situ XRD, in situ XAFS, and in situ TEM and EELS techniques. Further facilitated by DFT calculations, we found that the excessive lithium ions residing in the transition metal layer actually enhance the robustness and flexibility of the local structure, which enables the O–O bonds to contract forming peroxide-species and the local structure to distort to relieve the energy of the whole system, thus stabilizing the local structure during oxygen redox process.

3. We have successfully proved that the reversibility of the oxygen redox in $Li_2RuO_3$ can be tuned via doping a certain amount of boron atoms into the layered structure, which varies the electronic structure of Ru-O system. The results show that after the doping of boron atoms, $Li_2RuO_3$ exhibits more capacity contributed by Ru redox, while the capacity corresponding to oxygen redox is reduced. As a result, the modified $Li_2RuO_3$ presents great cycling stability with more reversible oxygen redox. Specifically, $Li_2RuB_{0.15}O_3$ can deliver 219 mAh $g^{-1}$ in the first cycle and can be cycled to 100th cycle with the retention of 86.7%.
4. The substitution of $Fe^{3+}$ in cubic $Li_2TiO_3$ will trigger the activity of oxygen redox, though $Fe^{3+}$ did not participate in the redox process. We have successfully made clear the activation mechanism by introducing the *U-Δ* theory derived from Zaanen–Sawatzky–Allen model and expanding it to a general level in the range of both lithium-ion and lithium-oxygen systems. Though high-covalent theory can explain the origin of anionic redox to certain extent, our theory shows more superiority in the manner of specific, theoretical, general, and quantitative in uncovering the activation mechanism.

Beijing, P. R. China
September 2018

Prof. Dingguo Xia

# Acknowledgements

Special thanks are due to my supervisor, Prof. Dingguo Xia, for his instruction on the research work and preparing for this Ph.D. thesis. He is not only the tutor of my scientific research, but also the tutor of my life.

We also thanks to the help from our co-authors of the research work in this thesis, names in no particular order: Prof. Ziyu Wu, Prof. Xiaodong Han, Prof. Wangsheng Chu, Dr. Weifeng Huang, Dr. Ruiwen Shao, Dr. Ning Jiang, Dr. Huaifang Shang, Dr. Jin Ma, Dr. Hang Wei, Dr. Li An, Dr. Huijun Yan, Dr. Yuxuan Zuo, Dr. Bin Zhang. Thanks to the students in the laboratory like Dr. Zhen Yu, Dr. Guang Feng, Dr. Jin Song, Dr. Fanghua Ning. We appreciated the help from the staff of beamlines 1W1B, 4B7A, and 4B7B of the Beijing Synchrotron Radiation Facility (BSRF), beamline BL10B of the National Synchrotron Radiation Laboratory (NSRL), and beamline BL14W1 of the Shanghai Synchrotron Radiation Facility for the XAS experiments.

# Contents

# Chapter 1
# Introduction

**Biao Li and Dingguo Xia**

The issues related to energy and environment are attracting more and more attentions with the proposing of sustainable development concept in twenty-first century. The exhausting of fossil fuels leads to energy crisis and environmental pollution, which generates a huge challenge for the persistent development of human society. Therefore, exploiting and utilizing new clean energy are of great significance in scientific research. As a crucial part of the energy storage materials, lithium-ion batteries play the most important role in the utilization of new energies, such as electric vehicles and large-scale energy storage. In China, the Ministry of Science and Technology issued New Energy Project for Electric Vehicle of National Key Research to promote the scientific research of power batteries, among which Li-ion battery is the predominant technology. Besides, the application of Li-ion battery in portable electronic devices, such as laptop, cell phone, and digital camera, is highly universal nowadays. Moreover, with the development of advance technologies like artificial intelligence, more and more mobile and intelligentialized electronic devices, e.g., drones, robots, and walking machine, lead to surged demand for Li-ion battery technologies. Hence, there is no doubt that developing Li-ion battery technology is meaningful for sustainable development in the future.

---

This chapter was adapted from B. Li et al. *Advanced Materials* 2017, 29 (48), 1701054 with permission. Copyright 2017, Wiley-VCH.

---

B. Li · D. Xia (✉)
Beijing Key Lab of Theory and Technology for Advanced Batteries Materials, College of Engineering, Peking University, Beijing 100871, P. R. China
e-mail: dgxia@pku.edu.cn

B. Li
e-mail: libiao2008@pku.edu.cn

 
B. Li, *Studies on Anionic Redox in Li-Rich Cathode Materials of Li-Ion Batteries*, Springer Theses, https://doi.org/10.1007/978-981-13-2847-3_1

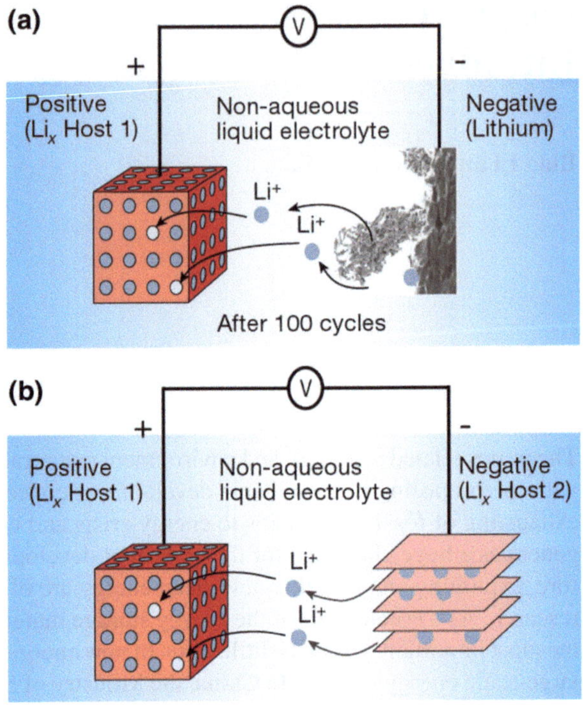


**Fig. 1.1** **a** Schematic of the basic structure of lithium battery and **b** schematic of the basic structure of lithium-ion battery. Reproduced with permission [3]. Copyright 2001, Springer Nature

## 1.1 Fundamental Principle of Li-Ion Batteries

In 1970s, M. S. Whittingham from Exxon company fist used lithium-intercalation compound $TS_2$ as the cathode electrode, lithium metal as the anode, fabricating the first lithium battery [1]. $TiS_2$ is a perfect layered compound, in which lithium can be reversibly intercalated/deintercalated. However, Li ions will be deposited on the metal anode forming lithium dendrite due to the inhomogeneous surface of lithium metal [2, 3], as demonstrated in Fig. 1.1a. The formation of lithium dendrite may penetrate the separator and thus cause short circuit, leading to fire and even explosion [2]. To overcome this problem, Armand et al. first proposed to use the lithium intercalated anode materials to substitute lithium [3], which was later realized by Murphy [4] and Scrosati et al. [5] that forms what we called lithium-ion batteries today or called as "rocking chair battery"; see Fig. 1.1b.

### *1.1.1 General Process for Redox Reactions*

An entire lithium rechargeable battery system usually consists of the cathode, anode, electrolyte, and separator. The cathode and anode are crucial to the battery performance and are the most studied components. The basic operating principle of the lithium-ion battery can be described as follows: Upon charging, the lithium ions deintercalate from the host (cathode) under the electric field and move to the anode host via the $Li^+$ conductive electrolyte; electrons are simultaneously donated by the cathode host, in order to maintain electric neutrality, and flow to the anode via an external electrically conductive circuit; the electrons form current and do work. The lithium deintercalation and electron donation cause redox of the electrodes, inducing electronic structure changes in the redox center.

In conventional views, the redox center in lithium-based metal oxides, sulfides, and polyanion-type compounds is usually considered to be the metal sites since the electrons in the metal d states are more active than those in the anion p states [6]. In most cases, the metal ions have multiple valence states, while the anions have invariable valence states. The common cathode material $LiFePO_4$ adopts an intercalation mechanism that exemplifies this, with the ferrous ion becoming trivalent upon delithiation, as expressed in Eq. (1.1) [7].

$$LiFe^{2+}PO_4 \leftrightarrow xLi^+ + Fe^{3+}PO_4 + (1-x)LiFe^{2+}PO_4 + xe^- \tag{1.1}$$

Conversely, compounds that contain metal ions with a single valence state, such as $LiAlO_2$, are usually considered to be electrochemically inactive and cannot deliver capacity within a certain voltage range [6]. Other redox reactions besides the intercalation mechanism can also take place. Conversion reactions, which are mostly seen in the anode redox process, occur in most transition metal oxides and sulfides, such as CoO, $Co_3O_4$, and $CoS_2$ [8, 9]. In a conversion reaction, the metal ions are totally reduced to their metallic state during lithiation of the anode, with lithium oxide formed, as illustrated in Eq. (1.2).

$$Co^{2+}O + 2Li^+ + 2e^- \rightarrow Li_2O + Co(s) \tag{1.2}$$

Another interesting redox process occurs when lithium ions intercalate into a carbon anode [10], as expressed in Eq. (1.3).

$$xLi^+ + C_6 + xe^- \rightarrow Li_xC_6 \tag{1.3}$$

The electrons from the cathode host are mostly distributed into the intercalated lithium ions forming metallic-like Li atoms, while the electronic structure of the carbon is also perturbed slightly. This reaction process is extremely close to that of the lithium metal anode, which is reflected by the low intercalation voltage that is close to zero (0.1 V) [11].

However, redox processes involving anions in lithium compounds are also possible. Regardless of the reversibility, it is theoretically possible to deintercalate lithium ions from cation–inert lithium compounds if a sufficiently high voltage is provided. Additionally, anions are also involved in the redox reactions of some lithium compounds with multivalent cations due to the unique electronic and geometric structure of these materials. Issues of intrigue concerning the activation of anion redox in lithium rechargeable batteries will be discussed and summarized later.

### 1.1.2 Band Structures for Redox Reactions

The basic principle of redox processes can be traced back to the band structure of the relevant cations and anions. Metal d states usually overlap with the ligand p states, forming M–L (M denotes transition metal, and L denotes the ligand) bonds with a certain ionic and covalent characteristics determined by the kind of metal and anions and also the crystal structure. Almost, all of the electrodes concerned in this review adopt a rock-salt structure [12], which comprises the metal and ligand atoms in octahedral coordination, as shown in Fig. 1.2a. Typically, the band structure of the σ overlap between transition metal and ligands in octahedral environment is illustrated detailedly in Fig. 1.2b. For the center M atom, its five nd orbitals are classified as $d_{x_2-y_2}$, $d_{z_2}$, $d_{xy}$, $d_{yz}$, and $d_{xz}$. Among those, $d_{x_2-y_2}$ and $d_{z_2}$ ($e_g$) will directly overlap with ligand p(σ) orbitals along the M–L bond directions to form antibonding $e_g^*$ band with higher energy and bonding $e_g$ band with comparatively lower energy. $d_{xy}$, $d_{yz}$, and $d_{xz}$ orbitals ($t_{2g}$) are commonly considered to form non-bonding states since they have no direct overlapping with p(σ) orbitals, regardless of little π overlapping. The d electrons of M are accommodated in these $e_g^*$ and $t_{2g}$ bands, which are labeled as M(d) states in Fig. 1.2b and stood out by sky blue color. Metal $(n+1)$s and $(n+1)$p states will overlap with ligand p(σ) states forming $a_{1g}$ and $t_{1u}$ bands. Due to the higher electronegativity and lower energy of the ligands compared with M, the electrons of ligand p states reside mostly in bonding states of, $e_g$, $a_{1g}$ and $t_{1u}$, which were generally intermixed with each other due to their similar energies and labeled together as L(p) states.

For a general delithiation redox process, the occupied states below the Fermi level will donate certain amounts of electrons corresponding to the number of extracted lithium ions, which is called a redox process. In a conventional case, it is M(d) states that are well separated from L(p) states (not plotted as Fig. 1.2b shown however) to give the M d electrons, resulting in the oxidation of M ions, thus acting as a cationic redox center. This is exemplified by the situation of $LiFePO_4$, where electrons are removed from Fe $t_{2g}$ states and $Fe^{2+}$ is oxidized to $Fe^{3+}$. However, if there is overlap between M(d) states and L(p) states as plotted in the figure (that L(p) area overlaps with M $t_{2g}$ band), electron transfer will also start to happen on ligand when Fermi level is pinned on the top of L(p) band during subsequent delithiation process. This will generate what we called as an anionic redox process, which is generally found in conventional layered oxides with high-covalent M–O bond, such as $LiCoO_2$.

However, mechanisms are not the same in lithium-rich electrodes, which will generate non-bonding ligand p states in a higher level than $LiMO_2$ (see the dashed ellipse area with yellow color between $t_{2g}$ and $e_g^*$ bands in Fig. 1.2b) due to the Li–O–Li configuration, as highlighted in Ceder's work previously [13]. It is noteworthy that the actual anionic redox processes are various in specific electrodes, given that the electronic structures are distinct in different compounds.

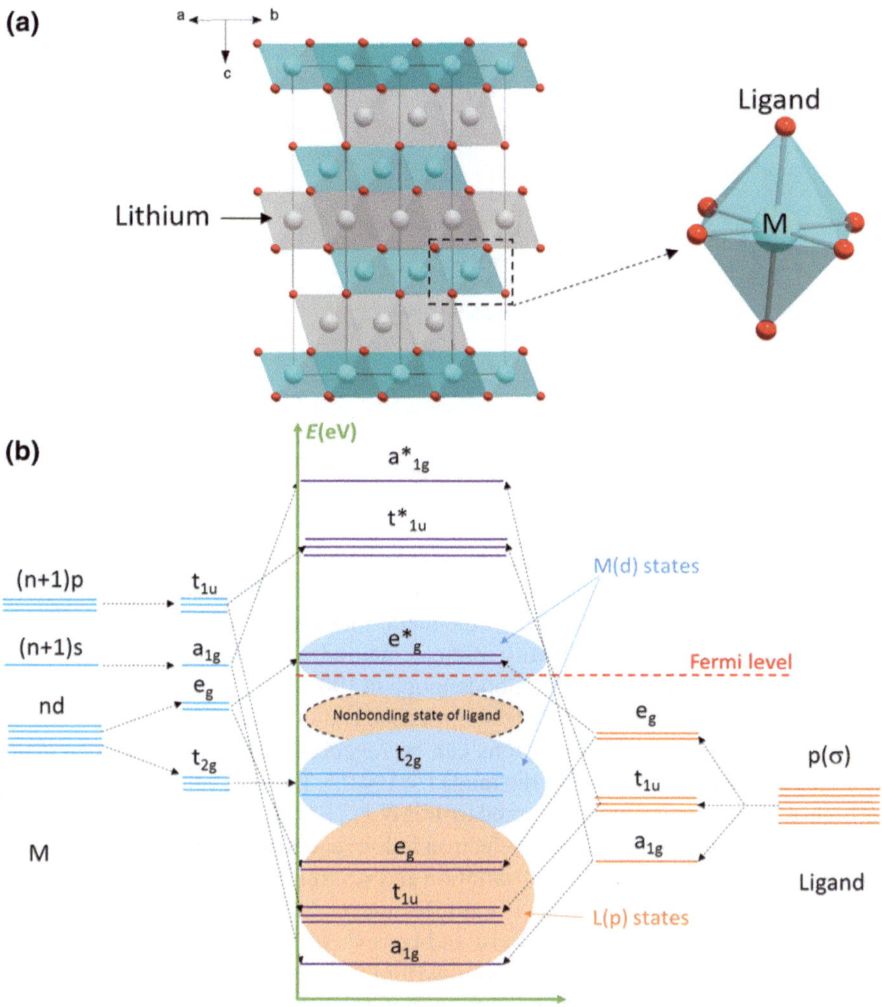


**Fig. 1.2** **a** Schematic of the crystal structure of lithium rock-salt oxides, with an enlarged ML6 octahedral and **b** band structures of the overlap between the M and ligand orbitals in a typical $ML_6$ coordination

## 1.2 Anionic Redox in Conventional Layered Oxide Electrodes

Many lithium hosts have been found and used as cathodes and anodes in lithium rechargeable batteries since M. S. Whittingham introduced the first lithium-intercalation concept materials in the 1970s; these include lithium rock-salt layered oxides [14], spinel oxides [15], and olivine compounds [7]; see Fig. 1.3 [16]. Among these materials, the rock-salt layered oxides are the most popular lithium-ion hosts

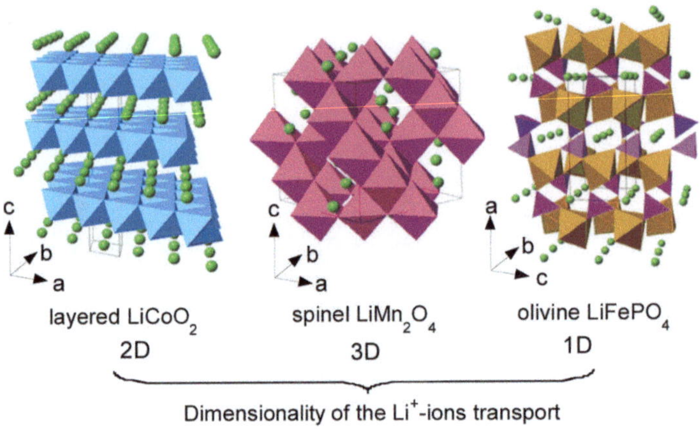


**Fig. 1.3** Three types of common cathode electrodes of Li-ion batteries

and comprise a large family based on different transition metal elements. Due to the d-p hybridization of transition metal and oxygen atoms, the oxygen electronic structure is influenced by variation of the metal ions during redox processes; hence, there is much debate on the identity of the actual redox center.

$LiCoO_2$, first proposed by J. B. Goodenough in 1980 [14], is the typical representative of layered rock-salt oxides and the most used cathode material in Li-ion batteries for portable electronics. Lithium ions can be reversibly intercalated/deintercalated from the host $Li_{1-x}CoO_2$ within the range $0<x\leq 0.5$. $LiCoO_2$ can deliver capacities as high as 150 mA h $g^{-1}$ when charged to 4.25 V [17]. However, its theoretical capacity is 272 mA h $g^{-1}$ based on a one-electron redox reaction in which lithium is totally extracted, but with large structural variation and several irreversible phase transitions that lead to quick capacity fading [17–19]. Though G. G. Amatucci et al. reported that the end product $CoO_2$ can stably exist and be reversibly inserted with lithium ions, $Li_{1-x}CoO_2$ can practically only be used within the range $0<x\leq 0.5$. Tremendous effort has been focused on enhancing the structural stability of $Li_{1-x}CoO_2$ at $x>0.5$, including doping and surface-coating modification [20–22]. However, the effects of such improvement have not been as good as was hoped, especially when $LiCoO_2$ is charged to a voltage greater than 4.5 V.

In order to determine the intrinsic origin of the irreversible structural change in $LiCoO_2$, much work has focused on probing the electronic structural change during charge and discharge. Many groups studied the redox reaction of $LiCoO_2$ both experimentally and theoretically to distinguish between the charge transfers on Co and O during delithiation. The redox reaction of $LiCoO_2$ remains controversial with three main ideas proposed: (1) $Co^{4+/3+}$ is traditionally thought to be the sole active component in $Li_{1-x}CoO_2$ redox [23]; (2) the electronic structure of Co does not obviously change during charging and discharging, with oxygen compensating most of the charge [24–29]; (3) both Co and O participate in the redox process, but contribute discriminately in different x ranges [7, 30–35].

Traditionally, taking $Co^{3+/4+}$ as the redox center of $LiCoO_2$ is the result of missing the band structural analysis, as shown in the introductory part. It is generally thought that M–O bonds in these rock-salt lithium-host oxides are of high iconicity, and thus has separate M d band and O 2p band. Therefore, electrons will only be considered to transfer from $t_{2g}$ band of Co 3d states, without any participation of oxygen. The valence change of Co ions has also been validated by Co K-edge X-ray absorption fine structure (XAFS) [23], though with no identification of how much the change is since the spectra shape changed so much and cannot be clearly distinguished.

Before the intensive debates on redox of $LiCoO_2$, the change of oxygen electronic structure was experimentally observed by O K-edge X-ray absorption spectroscopy (XAS) spectra in 1991, when J. van Elp et al. doped a different amount of lithium ions into CoO [36]. Though not tackling with redox issues in $Li_{1-x}CoO_2$, this work is a very important foresight of the role of oxygen in the process of charge compensation. The earliest work questions the Co-redox-center theory in $Li_{1-x}CoO_2$ came in 1997, when G. Ceder et al. found significant charge transfer is from oxygen in delithiating $LiMO_2$ oxides, however, by ab initio calculations and without any experimental observation [30]. It brought 10-year studies on the actual redox process of $Li_{1-x}CoO_2$ and its related conventional layered oxides.

It is particularly interesting that some findings show oxygen predominates the redox process rather than Co (the contribution of Co can be neglected) [24–29], since it totally overturns previous views and cannot be understood based on band structures. C. Wolverton and A. Zunger have proposed a “self-regulating response” theory to account for the constant charge around Co during delithiation via self-consistent calculations; see the charge-density difference map in Fig. 1.4 [29]. However, the charge difference maps were calculated without relaxation of the lattice parameters, which may have circumscribed rationality to simulate the practical process. The dynamic process of the change in Co–O hybridization is also invisible in their work. Based on theoretical band structures, Co is absolutely the first to donate its electrons, and its partial oxidation was also observed by a series of characterization techniques comprising hard and soft XAS, XPS, and magnetometry [23, 32–35]. The minor electronic structural changes of Co showed by other works [24–26] are possible due to experimental deviation, since $Co^{4+}$ is metastable and ex situ measurements would be difficult in capturing its existence.

The electronic structural origin of oxygen redox in $LiCoO_2$ is generally regarded as the high covalence between Co and O atoms. $Co^{3+}$ is normally in low spin state, and the remaining six 3d electrons all reside in non-bonding $t_{2g}$ ($d_{xy}$, $d_{yz}$, and $d_{xz}$) bands. Due to the high covalence of Co–O, Co 3d states overlap with O 2p states largely. Hence, electrons of Co 3d states ($t_{2g}$ band) will be first involved in redox process, while oxygen redox is triggered as soon as the Fermi level is pinned at the top of O 2p states. A metal-insulator transition of $Li_{1-x}CoO_2$ was observed in this process when $x > 0.3$ [37]. This phenomenon is typically induced by oxygen redox as its 2p states exceed the Fermi level. Another important thing here is that whether the removal of electrons from O 2p band is direct or indirect through a ligand-to-metal charge-transfer process. In fact, as early as 1999, Tarascon et al. have already foretold the formation of electron holes in anions due to the electronic transfer from anionic

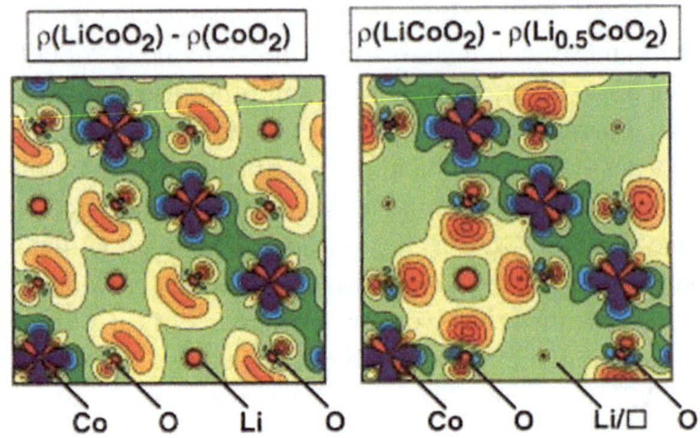


**Fig. 1.4** Charge-density difference map of $LiCoO_2$ before and after delithiation. Reproduced with permission [29]. Copyright 1998, American Physical Society

to d-cationic levels [38], after they found the shortening of interplanar O–O bond in delithiated $CoO_2$ and $NiO_2$ end members. The ligand-to-metal charge transfer means the charge redistribution between Co and O; that is, Co receives some electrons from oxygen and maintains less oxidized, while holes will be generated on oxygen. This very resembles the situation of the theoretical results obtained by C. Wolverton and Alex Zunger, who proposed the "self-regulating response" as previously referred. Since the charge redistribution may be transient, the electronic structural change of Co will be minor as experimentally proved by some groups, as referred before. Therefore, the issues associated with anionic redox in $LiCoO_2$ should be further investigated by other advanced techniques in the future. The high-penetrability and time-resolved operando synchrotron radiation characterization techniques may be very suitable to obtain such instant electronic structural information during the charge and discharge processes.

Other than $LiCoO_2$, other related conventional layered rock-salt oxides, such as $LiNiO_2$ [39–46], $LiNi_{0.8}Co_{0.2}O_2$ [47], $LiNi_{0.5}Mn_{0.5}O_2$ [48, 49], and $LiNi_{1/3}Co_{1/3}Mn_{1/3}O_2$ [50–54], were also reported with oxygen redox issues. This is not surprising since they all adopt similar oxygen networks and electronic structures. The same controversies like $LiCoO_2$ exist in identifying the redox in these oxides.

Unlike $Co^{3+}$, $Ni^{3+}$ has 7 d electrons, and the ground state of the Ni ions in $LiNiO_2$ and $LiCo_xNi_{1-x}O_2$ is commonly considered to be $Ni^{3+}$ ($t_{2g}^6e_g^1$). Based on this electronic state, $Ni^{3+}$ is expected to oxidize to $Ni^{4+}$ ($t_{2g}^6e_g^0$) upon delithiation without much involvement of oxygen 2p states. However, M. G. Kim et al. suggested that the initial electronic configuration of Ni ions exists as equilibrium between $Ni^{3+}\left(t_{2g}^6e_g^1\right) + O^{2-}(2p^6) \leftrightarrow Ni^{2+}\left(t_{2g}^6e_g^2\right) + O^{-}(2p^5)$ due to a ligand-to-metal charge-transfer (LMCT) process [39]. O K-edge XAS measurements showed that the $Ni^{3+}$–$O^{2-}$ states increase upon hole transfer from the O 2p to the Ni 3d states during delithiation, which can be interpreted as oxygen participating in the redox process accompanied by Ni ions. Another case is that Li/Ni disorder in $LiNiO_2$ will generate Li–O–Li configuration, leading to pure oxygen 2p band with higher energy [13], see Fig. 1.5, which finally induces oxygen redox.

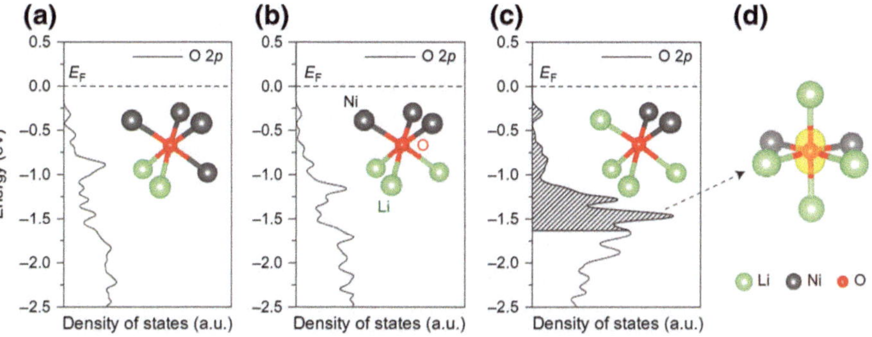


**Fig. 1.5** Analysis of the density of states of different local structures in Li/Ni-disordered $LiNiO_2$. Reproduced with permission [13]. Copyright © 2016, Springer Nature

Co-contained oxides are generally considered to involve oxygen 2p states in redox process due to the high-covalent Co–O bond. This has been validated by the comparison of the experimental results between Co-absent $LiNi_{0.5}Mn_{0.5}O_2$ and Co-contained $LiNi_{1/3}Co_{1/3}Mn_{1/3}O_2$ [48, 50]. This in some extent makes a misconception that the only origin of anionic redox is the high covalence between metal and anion, or interpreted as the large overlap between M(d) states with L(p) states, as explained in introduction part. However, this recognition was rectified after the appearance of lithium-rich rock-salt oxides, known for their high capacities. One typical lithium-rich oxide is $Li_2MnO_3$ and its derived oxides, which we will discuss next.

## 1.3 Anionic Redox in Li-Rich Oxide Electrodes

### *1.3.1 $Li_2MnO_3$ 与 $Li[Li_xMn_yNi_zCo_{1-x-y-z}]O_2$*

$Li_2MnO_3$ is the most studied of the model lithium-rich layered manganese oxide materials due to its peculiar electrochemical behavior; furthermore, the famously high-capacity $Li[Li_xMn_yNi_zCo_{1-x-y-z}]O_2$ is derived from it. The crystal structure of $Li_2MnO_3$ is based on the same cubic-close-packed oxygen networks as $LiMO_2$ (M = Ni, Co, Mn, etc.), while the symmetry reduces to *C2/m* from $R\bar{3}m$ as a certain amount of lithium atoms resides in the Mn layer, which results in a superlattice order but with a lower symmetry [55]. $Li_2MnO_3$ was initially deemed to be electrochemically inactive since $Mn^{4+}$ was not expected to oxidize to $Mn^{5+}$ or higher-valence states. However, a substantial capacity was delivered upon charging it beyond 4.5 V and lithium ions can also be reinserted, though with a large irreversible capacity loss [56–58]. These results have invoked a lot of debate on the electrochemical activity of $Li_2MnO_3$.

A large number of early studies on the $Li_2MnO_3$ delithiation mechanism showed the presence of $Li^+/H^+$ exchange upon delithiation through $^2H$ nuclear magnetic resonance (NMR) and neutron diffraction (ND) measurements [56, 57, 59–62]. $H^+$ is generated by oxidation of the non-aqueous electrolyte upon charging to a high voltage. Most of these studies used samples prepared by chemical delithiation or acid treatment, which form $H_xLi_{2-2\delta-x}MnO_{3-\delta}$ compounds with oxygen vacancies accompanied by $Li_2O$ dissolution. However, chemical delithiation is intrinsically different to electrochemical delithiation since there is no electron transfer without any oxidative/reductive reactions. This is especially true for $Li^+/H^+$ exchange, which is a topological reaction, though its feasibility was confirmed by free energy calculations [63]; no net charge is transferred between $Li^+$ and $H^+$ during the process so it cannot supply electrons to the external circuit. Therefore, the high first charge capacity of $Li_2MnO_3$ cannot simply be attributed to ion exchange; it happens but contributes nothing to the redox process.

P. G. Bruce et al. first proposed the oxygen loss mechanism to account for lithium extraction from $Li_2MnO_3$, accompanied by the simultaneous $Li^+/H^+$ exchange [57, 64, 65]. The combined removal of $Li^+$ and $O^{2-}$ proceeds as the formal removal of $Li_2O$. It was proposed that the remaining Mn ions migrate to fill the internal vacancies left in the transition metal layer upon consumption of the surface oxygen. However, Y.-S. Hong et al. measured the electrode weight during the first charge process of $Li_2MnO_3$-based oxide and found that the weight loss was not consistent with oxygen loss [66, 67]. Thus, the entire process cannot be explained solely by oxygen loss.

It was demonstrated that $Mn^{4+}$ does not obviously change during lithium extraction, with the charge-compensated by oxygen oxidation [68, 69]. Xiao et al. [69] calculated that electrons in the $Mn^{4+}$ 3d states have comparatively lower energies than those in the O 2p states. Thus, the charge transfer upon delithiation first involves electrons in the O 2p band; this was confirmed by the O 2p states dominating the density of state (DOS) beyond the Fermi level. Bader charge analysis also suggested the participation of oxygen, rather than Mn ions, in the redox process.

Though oxygen ions are oxidized in the delithiation process, it is not possible to confirm the removal of oxygen simply by determining the DOS and Bader charge evolution. This raises difficulties in determining whether the entire process is solely $O^{2-/-}$ or $O^{2-}/O_2$ redox, or a combination of both. Some indirect density functional theory (DFT) calculations introduced the oxygen vacancy formation enthalpy $\Delta H_O$, as defined in Eq. (1.4), to evaluate the oxygen stability in the lattice [70].

$$\Delta H_O(x) = E(Li_xMnO_{3-\delta}) + \delta/2E(O_2) - E(Li_xMnO_3), \tag{1.4}$$

The enthalpy exceeds zero when $x < 1$ in $Li_xMnO_3$, which implies oxygen gas release upon delithiation when $x$ is below 1, while $O^{2-/-}$ could account for the redox process when $1 < x < 2$; see Fig. 1.6. H. Chen et al. also showed that oxygen oxidation proceeds via $O^{2-}$–$O^-$ to $O_2{}^-$–$O_2$, as inferred by the progressively contracted O–O bonds from 2.75 to 1.21 Å [71], which appears more rational as the oxygen evolution must go through gradual steps. This evolution has been experimentally observed by O K-edge XAS, which proves the formation of peroxide and superoxide species, and

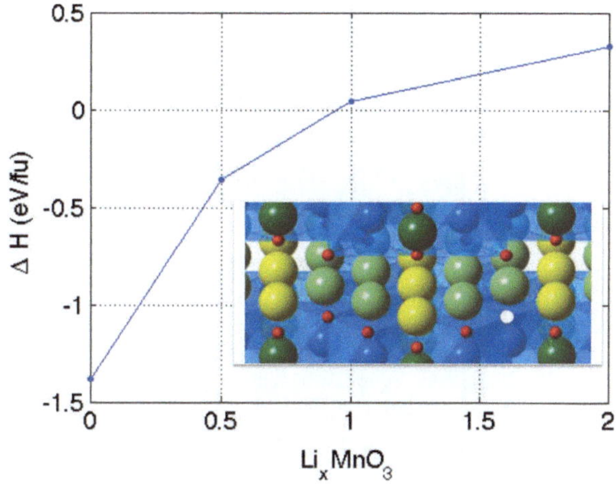


**Fig. 1.6** DFT calculated oxygen vacancy formation enthalpy in $Li_xMnO_3$. Reproduced with permission [70]. Copyright 2014, John Wiley and Sons

the oxygen redox occurs within the whole voltage range 2.0–4.8 V during charging/discharging, though $Mn^{4+/3+}$ is activated after the first charge within the lower voltage range [72]. Subsequent cycling of $Li_2MnO_3$ induces phase transition from layered to spinel, which cannot be primarily charge-compensated by oxygen as the typical spinel voltage dominates the charge/discharge curves. However, the kinetic favorability of the transformation was calculated, while oxygen redox was regarded as the main origin of the phase transition [73].

The above results demonstrate that oxygen redox and oxygen evolution are accounted for the whole lithium extraction in the first charge process of $Li_2MnO_3$. It should be noted that the ultimate purpose of intensive studies on $Li_2MnO_3$ is actually to understand related issues of its derived materials, $Li[Li_xMn_yNi_zCo_{1-x-y-z}]O_2$, which are famous for their reversible high capacities exceeding 250 mA h $g^{-1}$. Significant efforts were focused on determining the structure of $Li[Li_xNi_yCo_zMn_{1-x-y-z}]O_2$, which invoked a lot of debate on whether it is two-phase oxide that is a composite of $Li_2MnO_3$ [74–79] and $LiMO_2$ or a solid-solution [80–82], as shown in Fig. 1.7 (M equals Ni, Co, or Mn or their combination), phases [79]. Given the similarity of their electrochemical behavior, $Li[Li_xNi_yCo_zMn_{1-x-y-z}]O_2$ can be simply viewed as $Li_2MnO_3$ substituted by Ni or Co ions. The synergistic effect of the Ni and Co ions inhibits the phase transformation of $Li_2MnO_3$ after the first activation process and, thus, retains the initial layered structure.

Therefore, $Li[Li_xNi_yCo_zMn_{1-x-y-z}]O_2$ presents reversible high capacity. The identification of the origin of the excessive capacity of $Li[Li_xNi_yCo_zMn_{1-x-y-z}]O_2$ also went through the same controversy similar to $Li_2MnO_3$, including theories of $Li^+/H^+$ exchange [83, 84], oxygen loss (oxygen vacancies) [84–88], and reversible $O^{2-}/^{(2-n)-}$ redox [89–98]. Oxygen evolution rather than $O^{2-}/^{(2-n)-}$ redox at 4.5 V in the first charge process is fully accepted previously [87]. A model with different oxygen redox mechanisms in surface and bulk of $Li_{1.20}Mn_{0.54}Co_{0.13}Ni_{0.13}O_2$ was also

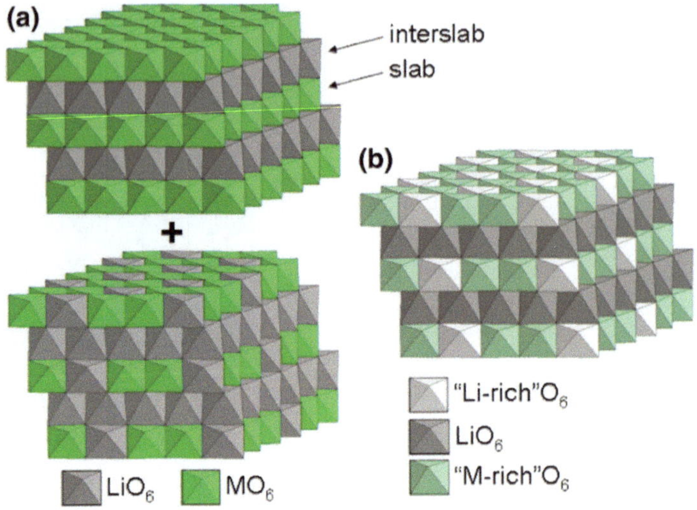


**Fig. 1.7** Schematic of **a** two-phase model and **b** solid-solution model of Li-rich Mn-based oxides. Reproduced with permission [79]. Copyright 2012, American Chemical Society

proposed by Koga et al. [92, 93]. They performed several studies and proposed that the surface oxygen is oxidized to molecular oxygen and then evolved, with the remaining metal ions migrating to fill the internal vacancy sites, while the bulk process consists of reversible oxygen redox without oxygen loss or structural rearrangement. To some extent, this mechanism is a reasonable explanation for some phenomena, such as the inconsistent weight loss and the Ni/Mn segregation in the surface and bulk [88, 94]. Besides, it can also explain the alteration of the charge behavior between the first cycle and subsequent cycles, since the surface change will influence the potential. Most recently, P. G. Bruce's group determined quantitatively that lithium extraction generating localized holes on the O 2p states coordinated by $Mn^{4+}$ and $Li^+$ ions predominates the 4.5 V process, with a small amount of oxygen loss, however, not as the form of oxygen gas [95, 96]. Their work clarified the key issue that oxygen redox generates electron holes rather than peroxide-like species, which was found by other groups in $Li_2MnO_3$ theoretically [71, 99]. Though peroxide-like species have been claimed to be found in $Li_{1.16}Ni_{0.15}Co_{0.19}Mn_{0.50}O_2$ and $Li_{1.14}Ni_{0.136}Co_{0.136}Mn_{0.544}O_2$ by XAS and XPS, respectively [97, 98], they have not differentiated whether the signals were caused by peroxide species or O 2p electron holes.

The redox process appears in $Li_2MnO_3$, and its derived oxides are confusing since there is no high covalence between Mn and O, as enlightened by $LiCoO_2$. Though previous work attributed oxygen redox to the Co ions in $Li[Li_xNi_yCo_zMn_{1-x-y-z}]O_2$, oxygen redox was also found in Co-free lithium-rich layered compound $Li[Li_{0.2}Ni_{0.2}Mn_{0.6}]O_2$ [96], which confirms that Co has no direct relationship with the oxygen redox. This indicates the origin of the anionic redox was not fully understood.

### *1.3.2 New Types of Li-Rich Oxide Electrodes*

Due to the success of $Li_2MnO_3$-based materials in delivering high capacities involving anionic redox, more lithium-rich layered oxides were proposed based on other lithium-rich metal compounds, such as $Li_2RuO_3$, $Li_2MoO_3$, and $Li_3NbO_4$. As expected, these layered materials mostly adopt the same rock-salt oxygen network like that of $LiCoO_2$ and $Li_2MnO_3$. All of these lithium-rich composites show comparatively high capacities compared with conventional layered oxides and involve multiple electron-transfer processes consisting of both cationic and anionic redox. In the following section, we review all of the M-based (M = Ti, Zr, Nb, Ru, Ir, Mo, Sb, etc.) lithium-rich layered compounds and their redox mechanisms.

#### 1.3.2.1 Ti-, Zr-, and Nb-Based Lithium-Rich Electrodes

Ti and Zr are similar to Mn in $Li_2MnO_3$ (Fig. 1.8), since they all form $Li_2MO_3$ (M = Ti and Zr) and cannot be further oxidized to higher valence in delithiation process. However, they have distinct electronic configurations because all of their d states are vacant without d electrons, which is also the situation of $Nb^{5+}$ in $Li_3NbO_4$. $Li_3NbO_4$ is a rock-salt-type oxides that is similar to $Li_2TiO_3$ and $Li_2ZrO_3$, but with more symmetry structure ($Fm\bar{3}m$) due to the disorder of Li and Nb ions (Fig. 1.8). The empty outmost d bands lead to very large band gaps and thus insulator properties in their corresponding oxides, which may have difficulty acting as electrode materials for lithium-ion batteries. Previous studies show expected results that $Li_2TiO_3$, $Li_2ZrO_3$, and $Li_3NbO_4$ all exhibited no electrochemical activity during charging and discharging processes [100–103].

However, if introducing active M ions into the structure of above insulative oxides, their activity may be activated. The cubic rock-salt lithium-rich structure $Li_2TiO_3$–$LiFeO_2$ was subsequently fabricated by Shigemura et al. [104]. This material has $Fm\bar{3}m$ symmetry, with disorderly Ti and Li atoms arranged at the cationic sites promoting the total symmetry compared with layered $Li_2TiO_3$. $Li_2TiO_3$–$LiFeO_2$ can deliver high capacities exceeding 200 mA h $g^{-1}$ in the first charge with a 4 V plateau, while there is a 3 V discharge plateau. In situ, $^{57}Fe$ Mossbauer spectra confirmed the presence of both $Fe^{4+/3+}$ and $Fe^{3+/2+}$ redox processes. Ti K-edge XAFS spectra demonstrated no obvious change in the Ti valence except in its local-range order. M. Tabuchi et al. also utilized in situ Mossbauer spectra and found the partial presence of $Fe^{4+/3+}$ in the first cycle, but did not find any evidence of $Fe^{3+/2+}$ in the 3 V discharge plateau [105]. By Rietveld refinement of X-ray diffraction (XRD) patterns, they proposed that oxygen loss occurs during the charging process, but they could not explain the redox mechanism of the 3 V discharge plateau. J. R. Dahn's group used XAS measurements to study these lithium titanium-based iron oxides [106]. They proposed that the redox mechanism of the 4.2 V charge process consists of $Fe^{4+/3+}$, $O^{2-/-}$, and oxygen loss, and the 3 V discharge process is charge-compensated by the combination of $Fe^{4+/3+}$, $O^{2-/-}$, and partial $Fe^{3+/2+}$, which is activated by the oxy-

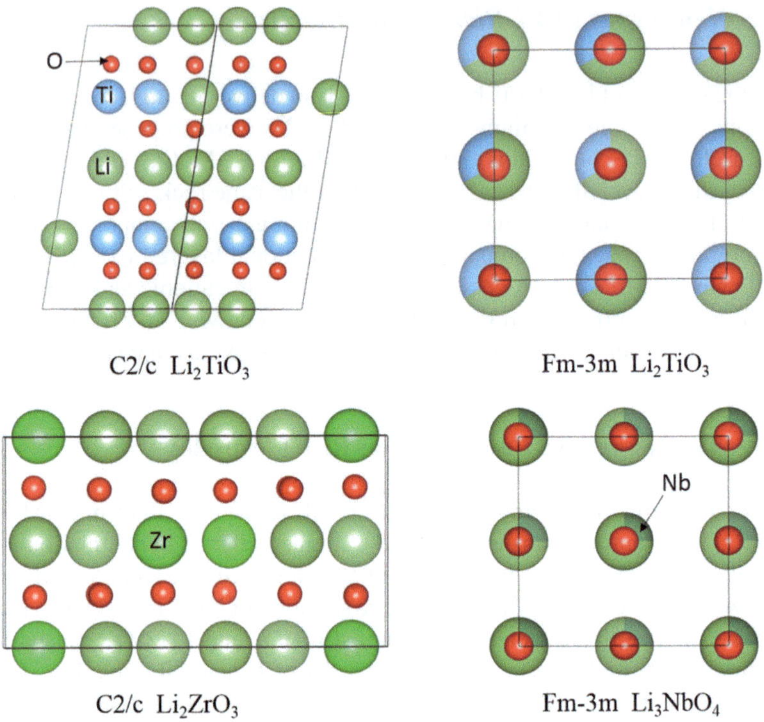


**Fig. 1.8** Structures of $Li_2TiO_3$, $Li_2ZrO_3$, and $Li_3NbO_4$

gen release. This peculiar electrochemical spectacle of large voltage hysteresis in Fe and O redox should be explored further as no consistent evidence was found by studying $LiFeO_2$ and other Fe-related titanium compounds [107, 108]. It should be noted that L. Zhang et al. studied a solid solution of $Li_2TiO_3$–$LiCrO_2$ and found similar 4 V charge and ~3 V discharge processes in $0.7Li_2TiO_3{\cdot}0.3LiCrO_2$ [109]. An abnormally high capacity at ~4 V was observed in the first cycle of the similar composite $LiNiO_2$–$Li_2TiO_3$ [110], which the authors attributed to the involvement of oxygen redox. Comparing all of these studies, the common peculiar electrochemical behavior appears to originate from Ti, with no Fe involvement. The inactive $Li_2TiO_3$ base determines the intrinsic charge/discharge properties, while the incorporation of active metal ions, such as Fe, Cr, and Ni, may trigger its electrochemical activity.

$Li_2ZrO_3$ was first proposed in $Li_2ZrO_3$–$LiMO_2$ (M = Ni, Co, Mn or their combination) composites for the stabilization of $LiMO_2$ oxides [102, 103]. No study has shown that $Li_2ZrO_3$ can be used as a base in designing lithium-rich oxides. However, another type of lithium-rich zirconium oxide with $R\bar{3}$ symmetry (Fig. 1.9a), $Li_8ZrO_6$, was examined experimentally and theoretically as a new cathode material in lithium-ion batteries [111]. DFT calculations predicted that the first lithium atom can be removed from $Li_8ZrO_6$ at an average voltage of 4.04 V. Lithium removal creates holes in the O 2p band bound to the lithium vacancy. The layered structure is

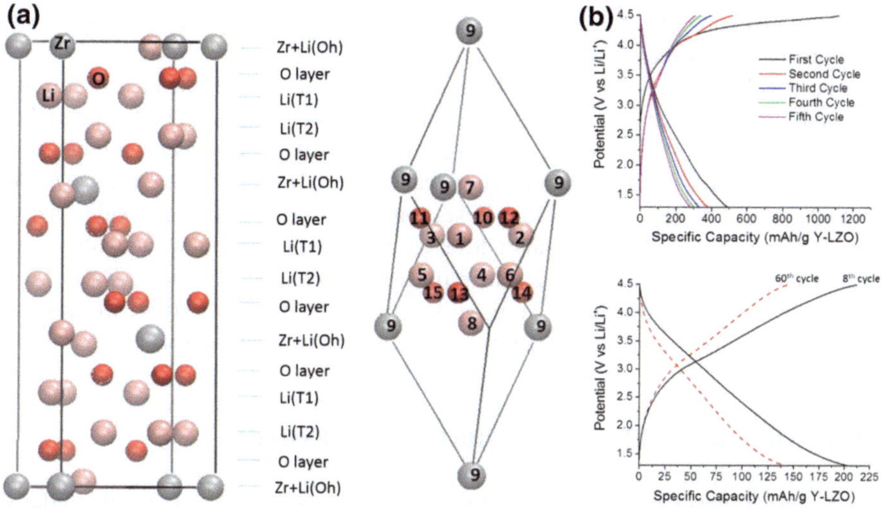


**Fig. 1.9** **a** Structure of $Li_8ZrO_6$ and **b** electrochemical performance of Y-$Li_8ZrO_6$

thermodynamically unstable when two or more lithium ions are removed, with oxygen gas released. Experimental studies showed that pure $Li_8ZrO_6$ can only deliver a small capacity of 20 mA h $g^{-1}$ due to its low electronic conductivity. The materials exhibit high capacities exceeding 500 mA h $g^{-1}$ after yttrium doping, but the capacity faded to about 130 mA h $g^{-1}$ after 60 cycles (Fig. 1.9b). Though the material demonstrates large hysteresis, the entire capacity originates from the oxygen redox, as proved by DFT calculations. However, the redox process was not studied by any experimental means.

When the molar ratio of cations/anions is kept at 1:1, lithium-rich layered oxides evolve from $LiM^{3+}O_2$ to $Li_2M^{4+}O_3$ to $Li_3M^{5+}O_4$, all adopting a NaCl-type rock-salt network. It is therefore Nb forms $Li_3NbO_4$ rather than $Li_2NbO_3$ since Nb has five outmost d electrons and usually is stable as $Nb^{5+}$. Thus, $Li_3NbO_4$ is electrochemically inactive except when some amount of active metal, like Mn, Ni, Co, Fe, is substituted for the Nb atoms. It has been shown that $Li_{1.3}Nb_{0.3}Mn_{0.4}O_2$ can exhibit reversible capacities of 300 mA h $g^{-1}$ at 50 °C for certain cycles [100]. Both experimental and theoretical studies indicate that the solid-state reaction of oxygen atoms is the primary charge compensation mechanism, in addition to the part contribution from Mn ions. This work demonstrates the feasibility of using $Li_3NbO_4$ as the base for designing new lithium-rich high-capacity electrodes involving both cationic and anionic redox processes. $Li_{1.25}Nb_{0.25}Mn_{0.5}O_2$ [100] with more manganese was later studied by R. Wang et al. and delivered a high capacity approaching 300 mA h $g^{-1}$ at 55 °C, with O K-edge EELS measurements proving the existence of oxygen redox. Substituting Nb with V atoms, as in $Li_{1.3}Nb_{0.3}V_{0.4}O_2$ [112], results in peculiar behavior. The material demonstrates capacities >200 mA h $g^{-1}$ without any oxygen redox; the charge is compensated solely by $V^{3+}/V^{5+}$ couples. This may be a result of the unique electronic structure of V ions, which can supply two electrons at comparatively lower voltages than other metal ions.

Since $Li_2TiO_3$ and $L_3NbO_4$ have been successfully utilized as the base of lithium-rich high-capacity oxides, the same expectation is also paid on $Li_2ZrO_3$ due to their similarity in electronic structure. However, there is little work concerning any doping of $Li_2ZrO_3$ to activate its electrochemical activity. If we consider further, $Fm\bar{3}m$ $Li_3TaO_4$ should be also applicable to designing Ta-based lithium-rich oxides. Moreover, $Li_4WO_5$ and $Li_5ReO_6$ are also deserved to be exploited, as well as other non-rock-salt lithium-rich oxides, which are represented by $Li_4Mn_2O_5$ and $Li_8ZrO_6$. From this perspective, it seems that a lot of work needs to be done to fertilize the cases of anionic redox, for the purpose of further recognizing its mechanism.

#### 1.3.2.2 Ru-, Ir-, and Mo-Based Lithium-Rich Electrodes

Ru and Ir elements have similar properties with Mn based on the diagonal similarity principle of periodic table of elements. Therefore, $Li_2RuO_3$ and $Li_2IrO_3$ can form nearly isostructural with *C2/m* $Li_2MnO_3$. However, the former two oxides have distinct redox processes upon delithiation compared with $Li_2MnO_3$, since $Ru^{4+}$ and $Ir^{4+}$ can more easily be oxidized to higher valence than $Mn^{4+}$. Though Ru and Ir are noble metals and not suitable for practical use, the distinct characteristics of Ru and Ir make $Li_2RuO_3$ and $Li_2IrO_3$ perfect models for mechanism studies of lithium-rich layered oxides. Tarascon et al. used a series of $Li_2M_{1-y}M'_yO_3$ (M = Ru and Ir; M′ = Mn, Sn, Ti) materials to study the redox processes and voltage decay of lithium-rich layered electrodes [113–115]. Ru-based lithium-rich materials are usually stable and deliver high capacities exceeding 200 mA h $g^{-1}$ over a 2.0–4.6 V voltage range, which involves both Ru and oxygen redox. The reversible oxygen redox in Sn-substituted $Li_2RuO_3$ was characterized by XPS and electron paramagnetic resonance (EPR) characterization, as shown in Fig. 1.10 [113]. A reductive coupling mechanism was proposed to explain the reduction of Ru ions upon charging to higher voltages [99]. This is discussed further in Sect. 1.6 along with the similar reduction of Ni ions in lithium-rich manganese-based materials.

Layered α-$Li_2IrO_3$ is theoretically calculated to be a conductor since Ir–O is more covalent than Ru–O [116]. Thus, the oxygen redox in α-$Li_2IrO_3$ is much expected based on previous outcomes. The higher covalence Ir–O than Co–O and Ru–O may generate more holes in the O 2p band upon lithium extraction. Tarascon's group has presented the first visualization of the O–O dimers in α-$Li_2IrO_3$-based materials using scanning transmission electron microscopy (STEM) [117]. This is of great significance since it provides optical clues for anionic redox for the first time. After that, they also found the β-$Li_2IrO_3$ with a tridimensional (3D) framework exhibits highly reversible oxygen redox activity, which fertilizes the geometrical types of electrodes concerning anionic redox [118].

Unlike $Li_2RuO_3$ and $Li_2IrO_3$, $Li_2MoO_3$ has a disordered α-$NaFeO_2$ structure with $R\bar{3}m$ symmetry and, notably, the Mo forms local $Mo_3O_{13}$ clusters [119]. Mo ions tend to migrate to the lithium layers when lithium ions are extracted. $Li_2MoO_3$ was

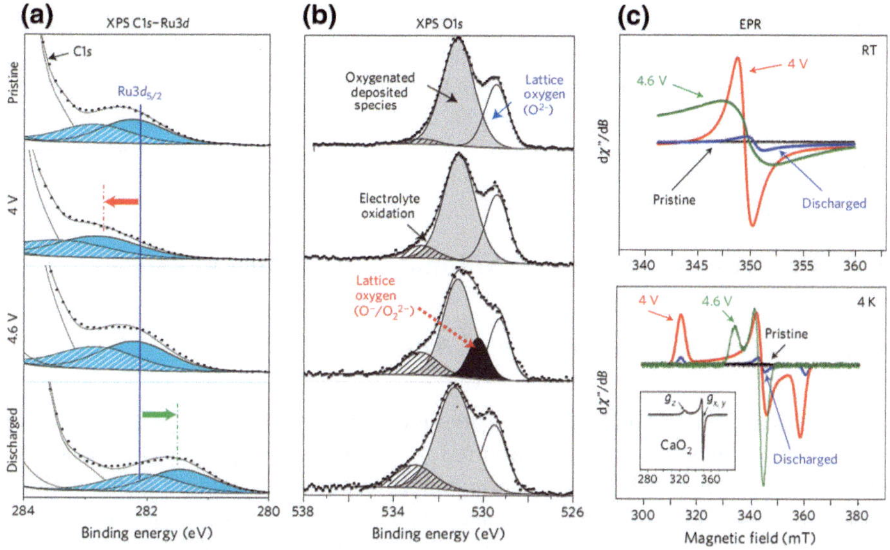


**Fig. 1.10** **a** C1s-Ru3d XPS peaks of $Li_2Ru_{0.5}Sn_{0.5}O_3$ during charging and discharging. **b** O 1s XPS peaks of $Li_2Ru_{0.5}Sn_{0.5}O_3$ during charging and discharging. **c** X band EPR spectra recorded at room temperature (top) and 4 K (bottom). Reproduced with permission [113]. Copyright © 2013, Springer Nature

studied in the hope of replacing $Li_2MnO_3$, though it was not stable for a long time when exposed to air. Pure $Li_2MoO_3$ utilizes partial $Mo^{6+/4+}$ and $O^{2-/(2-\delta)-}$ redox and did not display outstanding electrochemical performance, with poor stability and large hysteresis. J. Ma et al. evaluated the feasibility of using $Li_2MoO_3$ to construct Mo-based lithium-rich layered cathodes [120, 121], based on reducing the oxygen loss by increasing the electron transfer on the Mo sites. The Ceder group found that $Li_{1.211}Mo_{0.467}Cr_{0.3}O_2$ can also display high capacities exceeding 200 mA h $g^{-1}$, though cation disorder was present after several cycles, unlocking the potential for designing lithium-excess-disordered cathodes [122]. Though a high capacity corresponding to more one lithium atom per formula was delivered, the authors did not study the redox process in detail, in particular whether or not it involves oxygen redox. Ceder's group also sequentially prepared similar Li–Ni–Ti–Mo oxides that delivered capacities up to 250 mA h $g^{-1}$ [123]. The initial charging is compensated by $Ni^{2+/3+}$ redox, with oxygen loss and oxygen oxidation occurring in the latter stages of charging. The occurrence of oxygen redox after oxygen loss is not congruous to previous reports and quite confusing. Furthermore, the combination of multiple metals results in an intricate electronic structure and, consequently, the cause of the oxygen loss and oxygen redox remains mysterious.

$Li_2RuO_3$, $Li_2IrO_3$, and $Li_2MoO_3$ are somewhat different with $Li_2MnO_3$, $Li_2TiO_3$, and $Li_3NbO_4$, since the former three also involve metal redox themselves. However, if we consider a half delithiated states, such as $LiRuO_3$ and $LiIrO_3$, the subsequent redox processes are considered to be similar among them, since, from the perspective of band structure, it is O 2p states that dominate the vicinity of Fermi level in $LiRuO_3$, $LiIrO_3$, $Li_2MnO_3$, $Li_2TiO_3$, and $Li_3NbO_4$. The resulted question is, among these oxides, why some of them have oxygen redox activities and some not, and what decides the potential of oxygen redox.

## 1.4 Progress on Studies of Anionic Redox Mechanism

### *1.4.1 Activation Mechanism of Anionic Redox*

The cation and anion band structures are determined by the metal and anion types, obeying the principle of anion–cation redox competition as proposed by J. Rouxel et al. in the 1990s [124]. For transition metals on the left-hand side of the periodic table with lower electronegativities, such as Ti, V, and Cr, the d states usually are located at energy levels far above the oxygen 2p states. This explains why oxygen redox is seldom found in oxides like $LiTiO_2$ and $LiVO_2$. On the other hand, transition metals on the right-hand side, such as Co, Ni, and Cu, have large electronegativities and form highly covalent bonds with oxygen, with delocalized d and p states. Consequently, these oxides display highly complex redox processes, as previously discussed for $LiCoO_2$, $LiNiO_2$, and even $LiCuO_2$. Though we are accustomed to describing the electronic structure using the model of coordination polyhedral built up around cations, anions are also very crucial in determining the band structure. The energy of the top of the anion p band increases in order of oxygen, sulfur, and selenium, and the same trend is seen in other anionic groups like the pnictogens and halogens. Hence, sulfides and selenides are more likely to involve anionic redox processes as the p bands have a larger overlap with the metal d band.

Actually, no matter whether the M ions are active or not, the oxygen redox will be initiated when the top of O 2p states reaches the Fermi level. This truly happens in $Li_{1-x}CoO_2$ after a certain amount of electrons is removed from Co 3d ($t_{2g}$) states and Fermi level is pinned on the top of O 2p band, and in $Li_2MnO_3$ without any involvement of $Mn^{4+}$ since O 2p electrons predominate the vicinity of Fermi level. However, there remain some other unresolved questions that need to be explained. Lithium-rich M compounds without any active metal ions, or with M ions of $d^0$ electronic configuration, like layered $Li_2TiO_3$ and $Li_2ZrO_3$, cannot deliver substantial capacity via oxygen redox like $Li_2MnO_3$ does. Their insulating nature may be one possible explanation that accounts for the inactivity of these materials. However, we compared the band gaps of some common electrodes and found that polyanion-type cathodes, such as $LiMnPO_4$ and $LiCoPO_4$, have a larger band gap (~4.2 eV) [125] than layered $Li_2TiO_3$ (3.3 eV) [126]. Since $LiMnPO_4$ and $LiCoPO_4$ are electrochemically active, electronic conductivity may be not the main cause of the inactivity of $Li_2TiO_3$. It should be noted that lithium ions may be extracted from $Li_2TiO_3$ via oxygen redox under higher voltages over 5 V. This indicates that there is probably an energy barrier

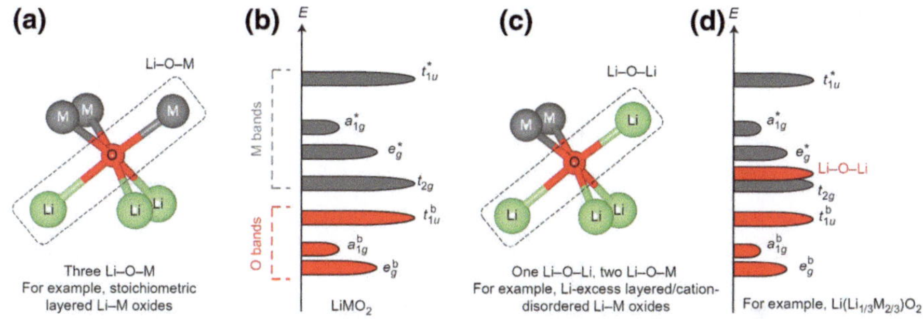


**Fig. 1.11** **a** Local coordination structure around Li–O–M configured M atoms in layered Li metal oxides. **b** Schematic of the band structure of $LiMO_2$ oxides. **c** Local coordination structure around M atoms with Li–O–Li and Li–O–M configurations in layered Li-excess metal oxides. **d** Schematic of the band structure of Li-excess layered oxides, such as $Li_2MO_3$. Reproduced with permission [13]. Copyright © 2016, Springer Nature

for the anionic redox activation, which is verified by the different activation voltages of oxygen redox in $Li_2MnO_3$, $Li_2RuO_3$, and $Li_2IrO_3$ [72, 113, 127].

Ceder et al. unraveled the chemical and structural origin of the oxygen redox in these lithium-rich layered oxides and cation-disordered oxides [13]. They found that oxygen redox is preferentially generated along the Li–O–Li configuration that is only present in lithium-excess or Li/M-disordered oxides and gives rise to labile electrons (generated by the non-bonding oxygen since Li 1s cannot overlap with O 2p, therefore Li–O–Li equals to O-non-bonding) in the O 2p band, as depicted in Fig. 1.11. This actually well explains the difference of redox mechanisms between conventional layered $LiMO_2$ and lithium-rich oxides, represented by $Li_2MO_3$ and $Li_3MO_4$. In ordered $LiMO_2$, oxygen ions are all coordinated to M ions and there are no pure non-bonding O 2p states. However, when Li/M increases associated with the change of the average valence of M (note here M may be a combination of the transition metals), Li–O–Li configurations will arise the pure non-bonding O 2p states with a high energy. Then, the oxygen redox in lithium-rich oxides is mainly contributed by this pure non-bonding 2p states, rather than the σ bonding states in the case of $LiMO_2$. With the increase of Li/M ratio, the configuration Li–O–Li increases and thus leads to greater contributions of oxygen redox, which is fully consistent with the increasing tendency of oxygen redox in $LiCoO_2$-, $Li_2RuO_3$-, and $Li_3NbO_4$-based oxides.

Cation-disordered lithium-excess oxides, such as $Li_3NbO_4$, can also be considered as layered structures, with lithium ions residing in every cationic layer. The flexibility of these structures is enhanced compared to ordered lithium-rich oxides, in which lithium ions are only present in the transition metal layer. A greater number of labile electrons will also arise from the increased Li–O–Li configuration in disordered lithium-rich oxides. From this perspective, the anionic redox may be more favorable in cubic disordered rock-salt-type lithium-rich oxides than in ordered ones.

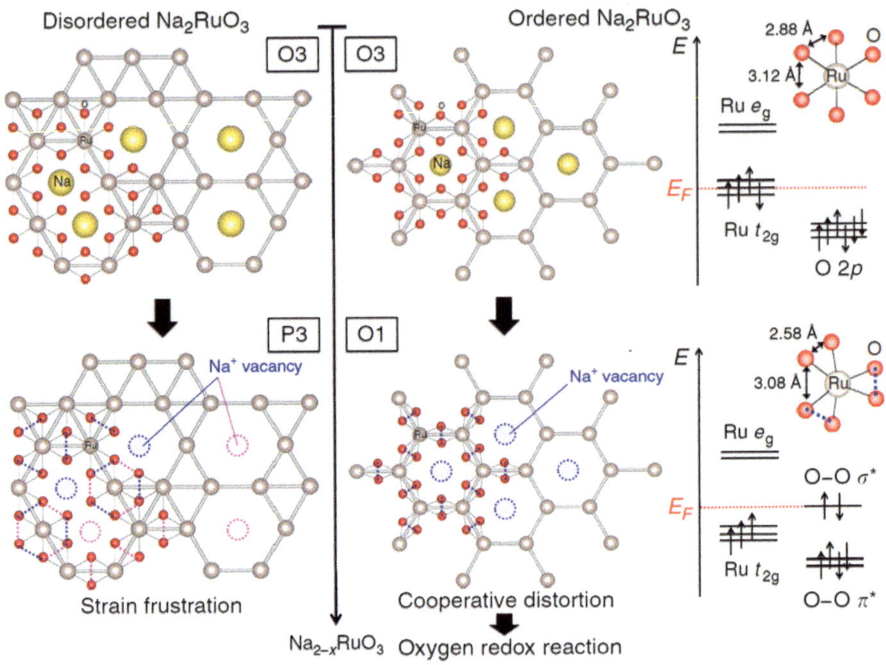


**Fig. 1.12** Superlattice ordering triggered oxygen redox in $Na_2RuO_3$

The superlattice order in the transition metal layer of $Na_2RuO_3$, used as a sodium-ion cathode, was also reported to be responsible for triggering anionic redox [128]. Ordered $Na_2RuO_3$ was found to accommodate the distortion induced by oxygen redox, while the disordered structure did not allow the distortion, which leads to strain frustration (Fig. 1.12). This further enhances the significance of lithium ions and their order in delivering high-capacity lithium-excess materials.

## *1.4.2 Reversibility of Anionic Redox*

In an anionic redox process, holes are first generated at the top of the anion p band and remain localized at specific anion sites. In a case of excessive anionic redox, the generated holes then condense to produce peroxides species and even oxygen gas, leading to oxygen removal from the lattice and phase transition of the crystal structure, as in the case of $LiCoO_2$ upon lithium extraction beyond 0.5 lithium ions per formula. Deep charge in the ternary oxide $LiNi_xCo_yMn_{1-x-y}O_2$ also leads to structural collapse and capacity fading due to the oxygen redox instability [129]. However, in lithium-rich manganese-based materials, lithium removal can be charge-compensated reversibly by the oxygen atoms, coordinated to the $Mn^{4+}$ and $Li^+$ ions, with localized electron holes formed on the O 2p band [130]. The high capacity can be maintained over a long cycle life, though further lithium removal causes oxygen loss, as observed at the end of the first 4.5 V voltage plateau. Therefore, it should be noted that the anionic redox stability plays a crucial role in delivering high capacity reversibly.

Tarascon et al. performed a series study on doped $Li_2RuO_3$ and characterized the oxygen redox reversibility [99]. Their work demonstrated that Ru ions are first oxidized upon lithium extraction and then reduced in the latter stages of charging. A reductive coupling mechanism (RCM) was proposed to explain this phenomenon, in which the high-lying oxygen 2p band electrons flow to the empty metal d band and thus lower the energy of the whole system; this was considered to be the primary factor that determines the oxygen redox reversibility. M.-L. Doublet et al. proposed that the metastable oxygen holes generated during delithiation are stabilized by the formation of $(O_2)^{n-}$ peroxo-like species via a M-driven RCM route. The absence of the RCM in $Li_2MnO_3$, in which there are no empty MO*-orbitals close enough to the top of oxygen 2p band to accommodate charge transfer from the oxygen, directly leads to oxygen evolution at the 4.5 V plateau. However, in Ni- and Co-substituted $Li_2MnO_3$, also a lithium-rich manganese-based oxide, Ni K-edge XAFS measurements identified an RCM in the Ni ion sites [131, 132]. This explains the reversibility of anionic redox in these lithium-rich layered oxides.

The irreversibility of oxygen redox in $LiCoO_2$ is rationalized by the absence of an RCM. However, it is still hard to explain why an RCM does not operate in the Co–O system, since its highly covalent nature is very similar to the Ru–O system. It is notable that RCMs have been found in lithium-rich oxides, but not in conventional layered oxides with the formula $LiMO_2$. This indicates that the triggering of an RCM may also be correlated with an excess of lithium ions. In reality, the geometrical structure is crucial in determining the stability of anionic redox. Actually, as stated by M.-L. Doublet et al., the oxygen redox in $LiMO_2$ consumes electrons from bonding states of the p bands and causes the structural collapse since these states are the fundamentals of the crystal structure integrity. While for lithium-rich oxides, the non-bonding states of O 2p band are the one to be consumed. The depletion of the electrons in the pure O 2p states without interaction with M ions has no direct damage to the structure, and it will also induce the formation of O–O peroxide species due to the RCM process.

Although an RCM can highly enhance the reversibility of anionic redox, the performance of oxygen redox over many cycles remains poor for most lithium-rich oxides. Actually, most of the studies on anionic redox almost focused on the first charge and discharge processes [91–93]. To fully assess the reversibility and stability of anionic redox, it is also important to consider subsequent cycles, especially for lithium-rich oxides which all exhibit distinct charge/discharge behaviors between the first cycle and subsequent cycles. However, little work has showed the accessibility of investigation on the subsequent cycles. R. Yuge et al. have unprecedentedly investigated the oxygen redox to 30 cycles in $Li_{1.26}Mn_{0.52}Fe_{0.22}O_2$, though the capacity deteriorates severely caused by redox irreversibility, including that of oxygen redox (Fig. 1.13) [133]. The reversibility of anionic redox thus has never been proved in

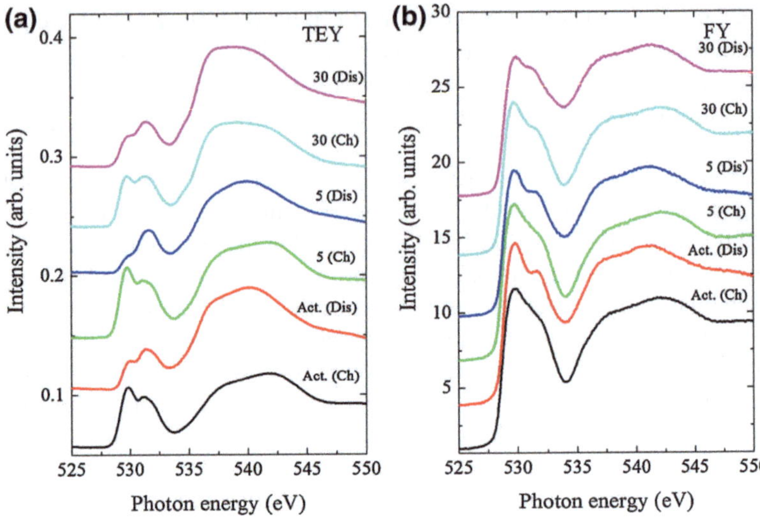


**Fig. 1.13** Oxygen redox in cycled $Li_{1.26}Mn_{0.52}Fe_{0.22}O_2$. Reproduced with permission [133]. 

longer cycles by any effective advanced characterization, except for extra-capacity expression caused surely by anionic redox. It is therefore important for future work on the characterization of reversibility of anionic redox in longer subsequent cycles. The participation of anionic redox should be limited to a certain range in order to obtain long cycle life. This presents a challenge for future electrode design in balancing the stability and high capacity.

## 1.5 The Purpose and Significance of the Topic

Redox process is the most fundamental process in Li-ion batteries and determines the electrochemical properties of the electrodes. The specific capacity of an electrode is determined by the electrons transferred during the electrochemical process; therefore, multiple electron-transfer processes are the key approach to realizing high capacity. Anionic redox provides more electron transfer based on cation redox, and thus more capacity, but also induces several problems, such as low columbic efficiency of the first cycle, oxygen release leading to safety concern, poor cyclability, large voltage hysteresis, and sluggish kinetics, which undermines the advantage of high capacity. Therefore, it is necessary to effectively modify the electrode materials based on the understanding of anionic redox. Besides, it is crucial to utilize anionic redox in designing high-capacity electrodes by intrinsically understanding its activation mechanism and reversibility.

## 1.6 Research Contents of This Thesis

Based on the deep understanding of anionic redox, we have tuned the redox process of lithium-rich electrode for achieving better electrochemical performance, uncovered the stability mechanism of the local structure upon anionic redox in lithium-rich cathode materials, as well as the activation mechanism of anionic redox in these electrode systems. The main contents include:

1. We have fabricated a kind of polyanion-modified lithium-rich manganese-based layered oxide, by introducing boracic polyanions into the layered structure to tune its electronic structure. The as-prepared electrode exhibits excellent cycling performance, specifically, showing 300 mA h $g^{-1}$ when cycled to 80th cycle under 0.1C rate current and 89% retention cycled to 300th cycle under 0.3C rate, with also higher thermal stability and charge/discharge potentials. Combined series of techniques, we have shown that the improvement should be ascribed to the incorporation of boracic polyanions, which lowers the overlap of M 3d (M = transition metal) and O 2p states, thus reducing the oxygen oxidation and resulting in a more stable structure.
2. Based on the model of $Li_2RuO_3$, the geometric and electronic structural evolutions of the lithium-rich high-capacity electrode were systematically investigated, with the aid of in situ XRD, in situ XAFS, and in situ TEM and EELS techniques. Further facilitated by DFT calculations, we found that the excessive lithium ions residing in the transition metal layer actually enhance the robustness and flexibility of the local structure, which enables the O–O bonds to contract forming peroxide species and the local structure to distort to relieve the energy of the whole system, thus stabilizing the local structure during oxygen redox process.
3. We have successfully proved that the reversibility of the oxygen redox in $Li_2RuO_3$ can be tuned via doping a certain amount of boron atoms into the layered structure, which varies the electronic structure of Ru–O system. The results show that after the doping of boron atoms, $Li_2RuO_3$ exhibits more capacity contributed by Ru redox, while the capacity corresponding to oxygen redox is reduced. As a result, the modified $Li_2RuO_3$ presents great cycling stability with more reversible oxygen redox. Specifically, $Li_2RuB_{0.15}O_3$ can deliver 219 mA h $g^{-1}$ in the first cycle and can be cycled to 100th cycle with the retention of 86.7%.
4. The substitution of $Fe^{3+}$ in cubic $Li_2TiO_3$ will trigger the activity of oxygen redox, though $Fe^{3+}$ did not participate in the redox process. We have successfully made clear the activation mechanism by introducing the $U$-$\Delta$ theory derived from Zaanen–Sawatzky–Allen model and expanding it to a general level in the range of both lithium-ion and lithium–oxygen systems. Though high-covalent theory can explain the origin of anionic redox in a certain extent, our theory shows more superiority in the manner of specific, theoretical, general, and quantitative in uncovering the activation mechanism.

## References


1. Whittingham MS. Electrical energy storage and intercalation chemistry. Science. 1976;192(4244):1126–7.
2. Armand MB. In materials for advanced batteries. In: Edsmurphy DW, Broadhead J, Steele BCH, editors. Proceedings of nato symposium on materials for advanced batteries; 1980. p. 145–61 (plenum, New york).
3. Tarascon JM, Armand M. Issues and challenges facing rechargeable lithium batteries. Nature. 2001;414(6861):359–67.
4. Murphy DW, Di Salvo FJ, Carides JN, et al. Topochemical reactions of rutile related structures with lithium. Mater Res Bull. 1978;13(12):1395–402.
5. Lazzari M, Scrosati B. A cyclable lithium organic electrolyte cell based on two intercalation electrodes. J Electrochem Soc. 1980;127(3):773–4.
6. Ceder G, Chiang YM, Sadoway DR, et al. Identification of cathode materials for lithium batteries guided by first-principles calculations. Nature. 1998;392(6677):694–6.
7. Padhi AK, Nanjundaswamy KS, Goodenough JB. Phospho-olivines as positive-electrode materials for rechargeable lithium batteries. J Electrochem Soc. 1997;144(4):1188–94.
8. Poizot P, Laruelle S, Grugeon S, et al. Nano-sized transition-metal oxides as negative-electrode materials for lithium-ion batteries. Nature. 2000;407(6803):496–9.
9. Yan JM, Huang HZ, Zhang J, et al. A study of novel anode material $CoS_2$ for lithium ion battery. J Power Sources. 2005;146(1–2):264–9.
10. Peled E, Menachem C, Bar-Tow D, et al. Improved graphite anode for lithium-ion batteries chemically: bonded solid electrolyte interface and nanochannel formation. J Electrochem Soc. 1996;143(1):L4–7.
11. Dahn JR, Zheng T, Liu Y, et al. Mechanisms for lithium insertion in carbonaceous materials. Science. 1995;270(5236):590.
12. Li B, Xia D. Anonic redox in lithium rechargeable batteries (submitted).
13. Seo DH, Lee J, Urban A, et al. The structural and chemical origin of the oxygen redox activity in layered and cation-disordered Li-excess cathode materials. Nat Chem. 2016;8(7):692–7.
14. Mizushima K, Jones PC, Wiseman PJ, et al. $Li_xCoO_2$ ($0<x\leq 1$): A new cathode material for batteries of high energy density. Mater Res Bull. 1980;15(6):783–9.
15. Thackeray MM, David WIF, Bruce PG, et al. Lithium insertion into manganese spinels. Mater Res Bull. 1983;18(4):461–72.
16. Julien C, Mauger A, Zaghib K, et al. Comparative issues of cathode materials for Li-ion batteries. Inorganics. 2014;2(1):132.
17. Ohzuku T, Ueda A. Solid-state redox reactions of $LiCoO_2$ ($R\bar{3}m$) for 4 volt secondary lithium cells. J Electrochem Soc. 1994;141(11):2972–7.
18. Reimers JN, Dahn JR. Electrochemical and in situ x-ray diffraction studies of lithium intercalation in $Li_xCoO_2$. J Electrochem Soc. 1992;139(8):2091–7.
19. Van der Ven A, Aydinol MK, Ceder G, et al. First-principles investigation of phase stability in $Li_xCoO_2$. Phys Rev B. 1998;58(6):2975–87.
20. Cho J, Kim YJ, Park B. Novel $LiCoO_2$ cathode material with $Al_2O_3$ coating for a Li ion cell. Chem Mater. 2000;12(12):3788–91.
21. Kim B, Lee JG, Choi M, et al. Correlation between local strain and cycle-life performance of $AlPO_4$-coated $LiCoO_2$ cathodes. J Power Sources. 2004;126(1–2):190–2.
22. Zou M, Yoshio M, Gopukumar S, et al. Synthesis of high-voltage (4.5 V) cycling doped $LiCoO_2$ for use in lithium rechargeable cells. Chem Mater. 2003;15(25):4699–702.
23. Nakai I, Takahashi K, Shiraishi Y, et al. Study of the jahn–teller distortion in $LiNiO_2$, a cathode material in a rechargeable lithium battery, by in situ x-ray absorption fine structure analysis. J Solid State Chem. 1998;140(1):145–8.
24. Montoro LA, Abbate M, Rosolen JM. Changes in the electronic structure of chemically deintercalated $LiCoO_2$. Electrochem Solid-State Lett. 2000;3(9):410–2.
25. Galakhov VR, Neumann M, Kellerman DG. Electronic structure of defective lithium cobaltites $Li_xCoO_2$. Appl Phys A. 2009;94(3):497–500.

26. Mesilov VV, Galakhov VR, Gizhevskii BA, et al. Charge states of cobalt ions in nanostructured lithium cobaltite: X-ray absorption and photoelectron spectra. Phys Solid State. 2013;55(5):943–8.
27. Koyama Y, Kim Yang-Soo, IsaoTanaka, et al. Changes in chemical bondings by Li deintercalation in $LiMO_2$ (M = Cr, V, Co and Ni). Jpn J Appl Phys. 1999;38(4R):2024.
28. Graetz J, Hightower A, Ahn CC, et al. Electronic structure of chemically-delithiated $LiCoO_2$ studied by electron energy-loss spectrometry. J Phys Chem B. 2002;106(6):1286–9.
29. Wolverton C, Zunger A. First-principles prediction of vacancy order-disorder and intercalation battery voltages in $Li_xCoO_2$. Phys Rev Lett. 1998;81(3):606–9.
30. Aydinol MK, Kohan AF, Ceder G, et al. Ab initio study of lithium intercalation in metal oxides and metal dichalcogenides. Phys Rev B. 1997;56(3):1354–65.
31. Seo DH, Urban A, Ceder G. Calibrating transition-metal energy levels and oxygen bands in first-principles calculations: accurate prediction of redox potentials and charge transfer in lithium transition-metal oxides. Phys Rev B. 2015;92(11):115118.
32. Yoon WS, Kim KB, Kim MG, et al. Oxygen contribution on Li-ion intercalation-deintercalation in $LiAl_yCo_{1-y}O_2$ investigated by O K-edge and Co L-edge X-ray absorption spectroscopy. J Electrochem Soc. 2002;149(10):A1305–9.
33. Yoon WS, Kim KB, Kim MG, et al. Oxygen contribution on Li-ion intercalation–deintercalation in $LiCoO_2$ investigated by O K-edge and Co L-edge X-ray absorption spectroscopy. J Phys Chem B. 2002;106(10):2526–32.
34. Mizokawa T, Wakisaka Y, Sudayama T, et al. Role of oxygen holes in $Li_xCoO_2$ revealed by soft X-ray spectroscopy. Phys Rev Lett. 2013;111(5):056404.
35. Klinser G, Topolovec S, Kren H, et al. Charging of lithium cobalt oxide battery cathodes studied by means of magnetometry. Solid State Ionics. 2016;293:64–71.
36. van Elp J, Wieland JL, Eskes H, et al. Electronic structure of coo, Li-doped CoO, and $LiCoO_2$. Phys Rev B. 1991;44(12):6090–103.
37. Menetrier M, Saadoune I, Levasseur S, et al. The insulator-metal transition upon lithium deintercalation from $LiCoO_2$: electronic properties and $^7Li$ nmr study. J Mater Chem. 1999;9(5):1135–40.
38. Tarascon JM, Vaughan G, Chabre Y, et al. In situ structural and electrochemical study of $Ni_{1-x}Co_xO_2$ metastable oxides prepared by soft chemistry. J Solid State Chem. 1999;147(1):410–20.
39. Kim MG, Sung NE, Shin HJ, et al. Ni and oxygen K-edge XAS investigation into the chemical bonding for lithiation of $Li_yNi_{1-x}Al_xO_2$ cathode material. Electrochim Acta. 2004;50(2–3):501–4.
40. Kuiper P, Kruizinga G, Ghijsen J, et al. Character of holes in $Li_xNi_{1-x}O$ and their magnetic behavior. Phys Rev Lett. 1989;62(2):221–4.
41. Abbate M, de Groot FMF, Fuggle JC, et al. Soft-x-ray-absorption studies of the location of extra charges induced by substitution in controlled-valence materials. Phys Rev B. 1991;44(11):5419–22.
42. van Elp J, Searle BG, Sawatzky GA, et al. Ligand hole induced symmetry mixing of $d^8$ states in $Li_xNi_{1-x}O$, as observed in Ni 2p x-ray absorption spectroscopy. Solid State Commun. 1991;80(1):67–71.
43. Mackrodt WC, Harrison NM, Saunders VR, et al. Direct evidence of O(p) holes in Li-doped NiO from Hartree-Fock calculations. Chem Phys Lett. 1996;250(1):66–70.
44. Chen H, Harding JH. Nature of the hole states in Li-doped NiO. Phys Rev B. 2012;85(11):115127.
45. Uchimoto Y, Sawada H, Yao T. Changes in electronic structure by Li ion deintercalation in $LiNiO_2$ from nickel L-edge and O K-edge XANES. J Power Sources. 2001;97–98:326–7.
46. Yukinori K, Isao T, Yang S, et al. First principles study on factors determining battery voltages of $LiMO_2$ (M = Ti–Ni). Jpn J Appl Phys. 1999;38(8R):4804.
47. Graetz J, Ahn CC, Yazami R, et al. An electron energy-loss spectrometry study of charge compensation in $LiNi_{0.8}Co_{0.2}O_2$. J Phys Chem B. 2003;107(13):2887–91.

48. Yoon W-S, Balasubramanian M, Yang X-Q, et al. Soft X-ray absorption spectroscopic study of a $LiNi_{0.5}Mn_{0.5}O_2$ cathode during charge. J Electrochem Soc. 2004;151(2):A246–51.
49. Miao S, Kocher M, Rez P, et al. Local electronic structure of layered $Li_xNi_{0.5}Mn_{0.5}O_2$ and $Li_xNi_{1/3}Mn_{1/3}Co_{1/3}O_2$. J Phys Chem B. 2005;109(49):23473–9.
50. Yoon W-S, Balasubramanian M, Chung KY, et al. Investigation of the charge compensation mechanism on the electrochemically Li-ion deintercalated $Li_{1-x}Ni_{1/3}Mn_{1/3}Co_{1/3}O_2$ electrode system by combination of soft and hard X-ray absorption spectroscopy. J Am Chem Soc. 2005;127(49):17479–87.
51. Tsai YW, Lee JF, Liu DG, et al. In-situ X-ray absorption spectroscopy investigations of a layered $LiNi_{0.65}Co_{0.25}Mn_{0.1}O_2$ cathode material for rechargeable lithium batteries. J Mater Chem. 2004;14(6):958–65.
52. Yoon W-S, Grey CP, Balasubramanian M, et al. Combined NMR and XAS study on local environments and electronic structures of electrochemically Li-ion deintercalated $Li_{1-x}Co_{1/3}Ni_{1/3}Mn_{1/3}O_2$ electrode system. Electrochem Solid-State Lett. 2004;7(3):A53–5.
53. Tsai YW, Hwang BJ, Ceder G, et al. In-situ X-ray absorption spectroscopic study on variation of electronic transitions and local structure of lini1/3co1/3mn1/3o2 cathode material during electrochemical cycling. Chem Mater. 2005;17(12):3191–9.
54. Petersburg CF, Li Z, Chernova NA, et al. Oxygen and transition metal involvement in the charge compensation mechanism of $LiNi_{1/3}Mn_{1/3}Co_{1/3}O_2$ cathodes. J Mater Chem. 2012;22(37):19993–20000.
55. Thackeray MM, Kang S-H, Johnson CS, et al. $Li_2MnO_3$-stabilized $LiMO_2$ (M = Mn, Ni, Co) electrodes for lithium-ion batteries. J Mater Chem. 2007;17(30):3112–25.
56. Robertson AD, Bruce PG. The origin of electrochemical activity in $Li_2MnO_3$. Chem Commun. 2002;23:2790–1.
57. Robertson AD, Bruce PG. Mechanism of electrochemical activity in $Li_2MnO_3$. Chem Mater. 2003;15(10):1984–92.
58. Croy JR, Park JS, Dogan F, et al. First-cycle evolution of local structure in electrochemically activated $Li_2MnO_3$. Chem Mater. 2014;26(24):7091–8.
59. Tang W, Kanoh H, Yang X, et al. Preparation of plate-form manganese oxide by selective lithium extraction from monoclinic $Li_2MnO_3$ under hydrothermal conditions. Chem Mater. 2000;12(11):3271–9.
60. Paik Y, Grey CP, Johnson CS, et al. Lithium and deuterium nmr studies of acid-leached layered lithium manganese oxides. Chem Mater. 2002;14(12):5109–15.
61. Venkatraman S, Manthiram A. Investigation of the possible incorporation of protons into oxide cathodes during chemical delithiation. J Solid State Chem. 2004;177(11):4244–50.
62. Lubin F, Lecerf A, Broussely M, et al. Chemical lithium extraction from manganese oxides for lithium rechargeable batteries. J Power Sources. 1991;34(2):161–73.
63. Benedek R, Thackeray MM, van de Walle A. Free energy for protonation reaction in lithium-ion battery cathode materials. Chem Mater. 2008;20(17):5485–90.
64. Armstrong AR, Bruce PG. Layered $Li_xMn_{1-y}Li_yO_2$ intercalation electrodes: synthesis, structure and electrochemistry. J Mater Chem. 2005;15(1):218–24.
65. Francis Amalraj S, Markovsky B, Sharon D, et al. Study of the electrochemical behavior of the "inactive" $Li_2MnO_3$. Electrochim Acta. 2012;78:32–9.
66. Hong Y-S, Park YJ, Wu X, et al. Synthesis and electrochemical properties of nanocrystalline $Li[Ni_{0.20}Li_{0.20}Mn_{0.60}]O_2$. Electrochem Solid-State Lett. 2003;6(8):A166–9.
67. Hong Y-S, Park YJ, Ryu KS, et al. Synthesis and electrochemical properties of nanocrystalline $Li[Ni_xLi_{(1-2x)/3}Mn_{(2-x)/3}]O_2$ prepared by a simple combustion method. J Mater Chem. 2004;14(9):1424–9.
68. Koyama Y, Tanaka I, Nagao M, et al. First-principles study on lithium removal from $Li_2MnO_3$. J Power Sources. 2009;189(1):798–801.
69. Xiao R, Li H, Chen L. Density functional investigation on $Li_2MnO_3$. Chem Mater. 2012;24(21):4242–51.
70. Lee E, Persson KA. Structural and chemical evolution of the layered Li-excess $Li_xMnO_3$ as a function of Li content from first-principles calculations. Adv Energy Mater. 2014;4(15):1400498.

71. Chen H, Islam MS. Lithium extraction mechanism in Li-rich $Li_2MnO_3$ involving oxygen hole formation and dimerization. Chem Mater. 2016;28:6656.
72. Oishi M, Yamanaka K, Watanabe I, et al. Direct observation of reversible oxygen anion redox reaction in Li-rich manganese oxide, $Li_2MnO_3$, studied by soft X-ray absorption spectroscopy. J Mater Chem A. 2016;4(23):9293–302.
73. Lim J-M, Kim D, Lim Y-G, et al. The origins and mechanism of phase transformation in bulk $Li_2MnO_3$: First-principles calculations and experimental studies. J Mater Chem A. 2015;3(13):7066–76.
74. Kang SH, Kempgens P, Greenbaum S, et al. Interpreting the structural and electrochemical complexity of 0.5 1 $Li_2MnO_3 \cdot 0.5$ $LiMO_2$ electrodes for lithium batteries (M = $Mn_{0.5-x}Ni_{0.5-x}Co_{2x}$, $0 \leq x \leq 0.5$). J Mater Chem. 2007;17(20):2069–77.
75. Lei CH, Bareño J, Wen JG, et al. Local structure and composition studies of $Li_{1.2}Ni_{0.2}Mn_{0.6}O_2$ by analytical electron microscopy. J Power Sources. 2008;178(1):422–33.
76. Bareno J, Lei CH, Wen JG, et al. Local structure of layered oxide electrode materials for lithium-ion batteries. Adv Mater. 2010;22(10):1122–7.
77. Mohanty D, Huq A, Payzant EA, et al. Neutron diffraction and magnetic susceptibility studies on a high-voltage $Li_{1.2}Mn_{0.55}Ni_{0.15}Co_{0.10}O_2$ lithium ion battery cathode: Insight into the crystal structure. Chem Mater. 2013;25(20):4064–70.
78. Yu H, Ishikawa R, So YG, et al. Direct atomic-resolution observation of two phases in the $Li_{1.2}Mn_{0.567}Ni_{0.166}Co_{0.067}O_2$ cathode material for lithium-ion batteries. Angew Chem Int Ed. 2013;52(23):5969–73.
79. Boulineau A, Simonin L, Colin J-F, et al. Evolutions of $Li_{1.2}Mn_{0.61}Ni_{0.18}Mg_{0.01}O_2$ during the initial charge/discharge cycle studied by advanced electron microscopy. Chem Mater. 2012;24(18):3558–66.
80. Lu Z, Beaulieu LY, Donaberger RA, et al. Synthesis, structure, and electrochemical behavior of $Li[Ni_xLi_{1/3-2x/3}Mn_{2/3-x/3}]O_2$. J Electrochem Soc. 2002;149(6):A778.
81. Jarvis KA, Deng Z, Allard LF, et al. Atomic structure of a lithium-rich layered oxide material for lithium-ion batteries: evidence of a solid solution. Chem Mater. 2011;23(16):3614–21.
82. Shukla AK, Ramasse QM, Ophus C, et al. Unravelling structural ambiguities in lithium- and manganese-rich transition metal oxides. Nat Commun. 2015;6:8711.
83. Armstrong AR, Robertson AD, Bruce PG. Overcharging manganese oxides: extracting lithium beyond $Mn^{4+}$. J Power Sources. 2005;146(1–2):275–80.
84. Robertson AD, Bruce PG. Overcapacity of $Li[Ni_xLi_{1/3-2x/3}Mn_{2/3-x/3}]O_2$ electrodes. Electrochem Solid-State Lett. 2004;7(9):A294.
85. Armstrong AR, Bruce PG. Electrochemistry beyond $Mn^{4+}$ in $Li_xMn_{1-y}Li_yO_2$. Electrochem Solid-State Lett. 2004;7(1):A1.
86. Lanz P, Sommer H, Schulz-Dobrick M, et al. Oxygen release from high-energy $xLi_2MnO_3 \cdot (1-x)LiMO_2$ (M = Mn, Ni, Co): electrochemical, differential electrochemical mass spectrometric, in situ pressure, and in situ temperature characterization. Electrochim Acta. 2013;93:114–9.
87. Armstrong AR, Holzapfel M, Novák P, et al. Demonstrating oxygen loss and associated structural reorganization in the lithium battery cathode $Li[Ni_{0.2}Li_{0.2}Mn_{0.6}]O_2$. J Am Chem Soc. 2006;128(26):8694–8.
88. Boulineau A, Simonin L, Colin JF, et al. First evidence of manganese-nickel segregation and densification upon cycling in Li-rich layered oxides for lithium batteries. Nano Lett. 2013;13(8):3857–63.
89. Hy S, Su W-N, Chen J-M, et al. Soft x-ray absorption spectroscopic and raman studies on $Li_{1.2}Ni_{0.2}Mn_{0.6}O_2$ for lithium-ion batteries. J Phys Chem C. 2012;116(48):25242–7.
90. Oishi M, Fujimoto T, Takanashi Y, et al. Charge compensation mechanisms in $Li_{1.16}Ni_{0.15}Co_{0.19}Mn_{0.50}O_2$ positive electrode material for Li-ion batteries analyzed by a combination of hard and soft X-ray absorption near edge structure. J Power Sources. 2013;222:45–51.
91. Koga H, Croguennec L, Ménétrier M, et al. Operando X-ray absorption study of the redox processes involved upon cycling of the Li-rich layered oxide $Li_{1.20}Mn_{0.54}Co_{0.13}Ni_{0.13}O_2$ in Li ion batteries. J Phys Chem C. 2014;118(11):5700–9.

92. Koga H, Croguennec L, Menetrier M, et al. Reversible oxygen participation to the redox processes revealed for $Li_{1.20}Mn_{0.54}Co_{0.13}Ni_{0.13}O_2$. J Electrochem Soc. 2013;160(6):A786–92.
93. Koga H, Croguennec L, Ménétrier M, et al. Different oxygen redox participation for bulk and surface: a possible global explanation for the cycling mechanism of $Li_{1.20}Mn_{0.54}Co_{0.13}Ni_{0.13}O_2$. J Power Sources. 2013;236:250–8.
94. Gu M, Belharouak I, Genc A, et al. Conflicting roles of nickel in controlling cathode performance in lithium ion batteries. Nano Lett. 2012;12(10):5186–91.
95. Luo K, Roberts MR, Hao R, et al. Charge-compensation in 3d-transition-metal-oxide intercalation cathodes through the generation of localized electron holes on oxygen. Nat Chem. 2016;8(7):684–91.
96. Luo K, Roberts MR, Guerrini N, et al. Anion redox chemistry in the cobalt free 3d transition metal oxide intercalation electrode $Li[Li_{0.2}Ni_{0.2}Mn_{0.6}]O_2$. J Am Chem Soc. 2016;138:11211.
97. Oishi M, Yogi C, Watanabe I, et al. Direct observation of reversible charge compensation by oxygen ion in Li-rich manganese layered oxide positive electrode material, $Li_{1.16}Ni_{0.15}Co_{0.19}Mn_{0.50}O_2$. J Power Sources. 2015;276:89–94.
98. Han S, Xia Y, Wei Z, et al. A comparative study on the oxidation state of lattice oxygen among $Li_{1.14}Ni_{0.136}Co_{0.136}Mn_{0.54}O_2$, $Li_2MnO_3$, $LiNi_{0.5}Co_{0.2}Mn_{0.3}O_2$ and $LiCoO_2$ for the initial charge–discharge. J Mater Chem A, 2015;3(22):11930–11939.
99. Saubanère M, McCalla E, Tarascon JM, et al. The intriguing question of anionic redox in high-energy density cathodes for Li-ion batteries. Energy Environ Sci. 2016;9(3):984–91.
100. Yabuuchi N, Takeuchi M, Nakayama M, et al. High-capacity electrode materials for rechargeable lithium batteries: $Li_3NbO_4$-based system with cation-disordered rocksalt structure. Proc Natl Acad Sci. 2015;112(25):7650–5.
101. Kim JS, Johnson CS, Thackeray MM. Layered $xLiMO_2{\cdot}(1-x)Li_2M'O_3$ electrodes for lithium batteries: a study of $0.95LiMn_{0.5}Ni_{0.5}O_2{\cdot}0.05Li_2TiO_3$. Electrochem Commun. 2002;4(3):205–9.
102. Johnson CS, Korte SD, Vaughey JT, et al. Structural and electrochemical analysis of layered compounds from $Li_2MnO_3$. J Power Sources. 1999;81–82:491–5.
103. Johnson CS, Kim J-S, Jeremy Kropf A, et al. The role of $Li_2MO_2$ structures (M = metal ion) in the electrochemistry of $xLiMn_{0.5}Ni_{0.5}O_2{\cdot}(1-x)Li_2TiO_3$ electrodes for lithium-ion batteries. Electrochem Commun. 2002;4(6):492–8.
104. Shigemura H, Tabuchi M, Sakaebe H, et al. Lithium extraction and insertion behavior of nanocrystalline $Li_2TiO_3$–$LiFeO_2$ solid solution with cubic rock salt structure. J Electrochem Soc. 2003;150(5):A638.
105. Tabuchi M, Nakashima A, Shigemura H, et al. Fine $Li_{(4-x)/3}Ti_{(2-2x)/3}Fe_xO_2$ $(0.18 \leq x \leq 0.67)$ powder with cubic rock-salt structure as a positive electrode material for rechargeable lithium batteries. J Mater Chem. 2003;13(7):1747.
106. Glazier SL, Li J, Zhou J, et al. Characterization of disordered $Li_{(1+x)}Ti_{2x}Fe_{(1-3x)}O_2$ as positive electrode materials in Li-ion batteries using percolation theory. Chem Mater. 2015;27(22):7751–6.
107. Kanno R, Shirane T, Kawamoto Y, et al. Synthesis, structure, and electrochemical properties of a new lithium iron oxide, $LiFeO_2$, with a corrugated layer structure. J Electrochem Soc. 1996;143(8):2435–42.
108. Ma S, Noguchi H. Electrochemical properties of $LiFe_5O_8$–$Li_4Ti_5O_{12}$ solid solution. J Electrochem Soc. 2001;148(6):A589.
109. Zhang L, Noguchi H. Novel layered Li–Cr–Ti–O cathode materials related to the $LiCrO_2$–$Li_2TiO_3$ solid solution. J Electrochem Soc. 2003;150(5):A601.
110. Zhang L, Wang X, Noguchi H, et al. Electrochemical and ex situ XRD investigations on $(1-x)LiNiO_2{\cdot}xLi_2TiO_3$ $(0.05 \leq x \leq 0.5)$. Electrochimia Acta 2004;49(20):3305–3311.
111. Huang S, Wilson BE, Wang B, et al. Y-doped $Li_8ZrO_6$: a Li-ion battery cathode material with high capacity. J Am Chem Soc. 2015;137(34):10992–1003.
112. Yabuuchi N, Takeuchi M, Komaba S, et al. Synthesis and electrochemical properties of $Li_{1.3}Nb_{0.3}V_{0.4}O_2$ as a positive electrode material for rechargeable lithium batteries. Chem Commun. 2016;52(10):2051–4.

113. Sathiya M, Rousse G, Ramesha K, et al. Reversible anionic redox chemistry in high-capacity layered-oxide electrodes. Nat Mater. 2013;12(9):827–35.
114. Sathiya M, Ramesha K, Rousse G, et al. High performance $Li_2Ru_{1-y}Mn_yO_3$ ($0.2 \leq y \leq 0.8$) cathode materials for rechargeable lithium-ion batteries: their understanding. Chem Mater. 2013;25(7):1121–31.
115. Sathiya M, Abakumov AM, Foix D, et al. Origin of voltage decay in high-capacity layered oxide electrodes. Nat Mater. 2015;14(2):230–8.
116. O'Malley MJ, Verweij H, Woodward PM. Structure and properties of ordered $Li_2IrO_3$ and $Li_2PtO_3$. J Solid State Chem. 2008;181(8):1803–9.
117. McCalla E, Abakumov AM, Saubanère M, et al. Visualization of O–O peroxo-like dimers in high-capacity layered oxides for Li-ion batteries. Science. 2015;350(6267):1516–21.
118. Paul E, Pearce AJP, Rousse G, et al. Evidence for anionic redox activity in a tridimensional-ordered Li-rich positive electrode $Li_2IrO_3$. Nat Mater 2017.
119. James ACWP, Goodenough JB. Structure and bonding in $Li_2MoO_3$ and $Li_{2-x}MoO_3$ ($0 \leq x \leq 1.7$). J Solid State Chem. 1988;76(1):87–96.
120. Ma J, Gao Y, Wang Z, et al. Structural and electrochemical stability of Li-rich layer structured $Li_2MoO_3$ in air. J Power Sources. 2014;258:314–20.
121. Ma J, Zhou Y-N, Gao Y, et al. Feasibility of using $Li_2MoO_3$ in constructing Li-rich high energy density cathode materials. Chem Mater. 2014;26(10):3256–62.
122. Lee J, Urban A, Li X, et al. Unlocking the potential of cation-disordered oxides for rechargeable lithium batteries. Science. 2014;343(6170):519–22.
123. Lee J, Seo D-H, Balasubramanian M, et al. A new class of high capacity cation-disordered oxides for rechargeable lithium batteries: Li–Ni–Ti–Mo oxides. Energy Environ Sci. 2015;8(11):3255–65.
124. Rouxel J. Comments about cationic-anionic redox competition in the solid state. The formation of anion associations in the solid state. Comments Inorg Chem. 1993;14(4):207–28.
125. Shang SL, Wang Y, Mei ZG, et al. Lattice dynamics, thermodynamics, and bonding strength of lithium-ion battery materials LiMPO4 (M = Mn, Fe Co, and Ni): a comparative first-principles study. J Mater Chem. 2012;22(3):1142–9.
126. Wan Z, Yu Y, Zhang HF, et al. First-principles study of electronic, dynamical and thermodynamic properties of $Li_2TiO_3$. Eur Phys J B 2012;85(6).
127. Kobayashi H, Kanno R, Tabuchi M, et al. Structure and charge/discharge characteristics of new layered oxides: $Li_{1.8}Ru_{0.6}Fe_{0.6}O_3$ and $Li_2IrO_3$. J Power Sources. 1997;68(2):686–91.
128. Mortemard de Boisse B, Liu G, Ma J, et al. Intermediate honeycomb ordering to trigger oxygen redox chemistry in layered battery electrode. Nat Commun. 2016;7:11397.
129. Jung S-K, Gwon H, Hong J, et al. Understanding the degradation mechanisms of $LiNi_{0.5}Co_{0.2}Mn_{0.3}O_2$ cathode material in lithium ion batteries. Adv Energy Mater 2014;4(1):1300787.
130. Wei GZ, Lu X, Ke FS, et al. Crystal habit-tuned nanoplate material of $Li[Li_{1/3-2x/3}Ni_xMn_{2/3-x/3}]O_2$ for high-rate performance lithium-ion batteries. Adv Mater. 2010;22(39):4364–7.
131. Li B, Yan H, Ma J, et al. Manipulating the electronic structure of Li-rich manganese-based oxide using polyanions: towards better electrochemical performance. Adv Func Mater. 2014;24(32):5112–8.
132. Yabuuchi N, Yoshii K, Myung ST, et al. Detailed studies of a high-capacity electrode material for rechargeable batteries, $Li_2MnO_3$-$LiCo_{1/3}Ni_{1/3}Mn_{1/3}O_2$. J Am Chem Soc. 2011;133(12):4404–19.
133. Yuge R, Toda A, Kuroshima S, et al. Charge compensation mechanism during cycles in Fe-containing $Li_2MnO_3$ cathode for high energy density and low-cost lithium-ion batteries. Electrochim Acta. 2016;189:166–74.

# Chapter 2
# Experimental Section

## 2.1 Reagents

See (Table 2.1).

**Table 2.1** Reagents used in experiments

| Reagent | Formula | Specification | Manufacturer |
|---|---|---|---|
| Lithium acetate | $C_2H_3O_2Li\cdot H_2O$ | Analytical purity | Sinopharm Co., Ltd. |
| Cobalt(II) acetate | $C_4H_6CoO_4i\cdot 4H_2O$ | Analytical purity | Sinopharm Co., Ltd. |
| Manganese acetate | $C_4H_6MnO_4i\cdot 4H_2O$ | Analytical purity | Sinopharm Co., Ltd. |
| Nickel(II) acetate | $C_4H_6NiO_4i\cdot 4H_2O$ | Analytical purity | Sinopharm Co., Ltd. |
| Ethylene glycol | $C_2H_6O_2$ | Analytical purity | Sinopharm Co., Ltd. |
| Citric acid | $C_6H_8O_7i\cdot H_2O$ | Analytical purity | Xilong Scientific |
| Boric acid | $H_3BO_3$ | 99.99% | J&K |
| Ruthenium dioxide | $RuO_2i\cdot xH_2O$ | 54% Ru | J&K |
| Lithium carbonate | $Li_2CO_3$ | Analytical purity | Sinopharm Co., Ltd. |
| Titanium sulfate | $Ti(SO_4)_2$ | CP, ≥96% | Sinopharm Co., Ltd. |
| Ferrous sulfate | $FeSO_4i\cdot 7H_2O$ | AR, ≥99.5% | Sinopharm Co., Ltd. |
| Potassium hydroxide | KOH | AR, ≥85.0% | Sinopharm Co., Ltd. |
| Lithium hydroxide | $LiOHi\cdot H_2O$ | AR, ≥95% | Sinopharm Co., Ltd. |
| PVDF | $-(CH_2-CF_2)_n-$ | Analytical purity | Solvay |
| Super P | C | Chemical purity | WILLEBROEK-BELGIUM |
| PTFE | $-(CF_2-CF_2)_n-$ | Chemical purity | DuPont |

(continued)



B. Li, *Studies on Anionic Redox in Li-Rich Cathode Materials of Li-Ion Batteries*,
Springer Theses, https://doi.org/10.1007/978-981-13-2847-3_2

**Table 2.1** (continued)

| Reagent | Formula | Specification | Manufacturer |
|---|---|---|---|
| NMP | $C_5H_9NO$ | Analytical purity | Tianjin Fuchen |
| Nitric acid | $HNO_3$ | Analytical purity | Sinopharm Chemical Reagent Co., Ltd. |
| Hydrochloric acid | HCl | Analytical purity | Sinopharm Chemical Reagent Co., Ltd. |
| DMC | $C_3H_6O_3$ | Analytical purity | Tianjin Fuchen |

## 2.2 Instruments

See (Table 2.2).

**Table 2.2** Instruments used in experiments

| Instruments | Type | Manufacturer |
|---|---|---|
| Inert-gas glove box | UNilab | M. Braun |
| Furnace | SK-G08123 K | Central Furnace |
| Oven | Venticell | MMM Group |
| Rotary evaporator | Hei-VAP Advantage | Heidolph |
| Ultrasonic cleaner | KQ-300VDE | Kunshan Ultrasonic |
| Magnetic stirrer | RT 10 | IKA |
| Vacuum oven | VOS-60A | STIK |
| TG | EXSTAR SII TG/DTA 7300 | S II |
| Electronic balance | XS205 | METTLER TOLEDO |
| Electric roller press | MR-100A | Hefei Kejing |
| Electrochemical workstation | Parstat Multichannel | AMETEK |
| Temperate box | KB53 | BINDER |
| Battery packaging machine | MSK-110 | Hefei Kejing |
| Battery tester | BTS-5V10 mA | Neware |
| Slicer | T-10 | Hefei Kejing |
| Ball mill | Pulverisette 7 | FRITSCH |
| Water pump | A-1000S | EYELA |
| Diaphragm pump | C410 | Chemvak |
| Centrifuge | TG18 M | Changsha pingfan |
| Homogenizer | T18 basic | IKA |

## 2.3 Characterizations

### *2.3.1 Characterization of the Structures and Properties of the Samples*

See (Table 2.3).

**Table 2.3** Characterization techniques

| Techniques | Type | Specifications and conditions |
|---|---|---|
| X-ray diffraction | Bruker D8 Advance | Cu Kα radiation (λ = 1.5406 Å) source, operated at 40 kV and 40 mA; 10–80°, with the step size was 0.02°. |
| Scanning electron microscope | Hitachi S-4800 | Cold field, acceleration voltage 0.5–30 kV, secondary electronic image resolution 1 nm. |
| Transmission electron microscope | FEI Tecnai F20 | Point resolution 0.24 nm, line resolution 0.102 nm, magnification times 25X-1050kX |
| X-ray photoelectron spectroscopy | Axis Ultra | Al Ka, hv = 1486.7 eV, 600 W, 15 kV, 40 mA; energy resolution 0.48 eV(Ag $3d^{5/2}$) |
| Inductively coupled plasma spectroscopy | PROFILE SPEC Leeman | Optical resolution: ≤0.007 nm; stability: ≤1.0%; repeatability ≤1.0%; detection limit: ppm; wavelength range: 165–1100 nm |
| Fourier transform infrared spectrometer | Nicolet ECTOR22 | Highest resolution: 0.09 $cm^{-1}$, Mid-infrared: 4000–400 $cm^{-1}$, Far infrared: 700–50 $cm^{-1}$, Near infrared: 11000–3800 $cm^{-1}$ |
| Energy spectrum analyzer | GENESIS XM2 60S | Cold field, working voltage 0.5–30 kV, highest resolution 1μm |
| Synchrotron radiation X-ray absorption spectrum | Beijing Synchrotron Radiation Facility 4B9B/1W1B beamline, Shanghai Synchrotron Radiation Facility BL14W1 beamline, and National Synchrotron Radiation Laboratory BL10B beamline | Si (111) double-crystal monochromator, XAS data analysis was carried out using standard procedures by IFEFFIT |
| Differential scanning calorimeter | METTLER TOLEDO | Heating rate is 10 °C $min^{-1}$, temperature range is room temperature to 400 °C |

### 2.3.2 Electrochemical Performance

| Method | Type of instruments | Cells |
|---|---|---|
| Galvanostatic charge-discharge | Nware CT-3008-5 V10 mA-S1cell tester | CR2032 coin cell |
| Cyclic voltammetry | AMETEK | CR2032 coin cell |
| Galvanostatic intermittent titration technique | AMETEK | CR2032 coin cell |

# Chapter 3
# Polyanion-Modified Li-Rich Manganese-Based Layered Materials

**Biao Li, Huijun Yan, Jin Ma, Pingrong Yu, Dingguo Xia, Weifeng Huang, Wangsheng Chu and Ziyu Wu**

## 3.1 Introduction

Owing to the rapid development of portable electronics, electric vehicles, and energy storage devices, the demand for lithium-ion batteries (LIBs) with high energy density and long cycle life has spurred enormous research efforts to develop novel electrode materials with high performance. The conventional layered transition metal oxide cathode materials, such as $LiCoO_2$, $LiNi_{1/3}Co_{1/3}Mn_{1/3}O_2$, and $LiNi_{0.8}Co_{0.2}O_2$, despite of their high theoretical capacities, suffer a drawback of oxygen evolution in high delithiated state, which could cause severe thermal runaway of the cell for practical application and lead to low practical capacity [1–4]. Lithium-rich manganese-based layered materials, $x\mathrm{Li_2MnO_3}\cdot(1-x)\mathrm{Li[Ni_{1/3}Co_{1/3}Mn_{1/3}]O_2}$, can deliver an anomalously high discharge capacity of more than 250 mAh $g^{-1}$ through the activation of $Li_2MnO_3$ and have attracted increasing attention as promising high-capacity cathode [5–9]. However, several drawbacks, including poor cycle life, low thermal stability, and a gradual voltage drop during cycling, hindered its commercial application in Li-ion batteries [10, 11]. To overcome these problems, tremendous efforts

This chapter was adapted from B. Li et al. *Advanced Functional Materials* 2014, 24(32), 5112–5118 with permission. Copyright 2014, Wiley-VCH.

B. Li · H. Yan · J. Ma · P. Yu · D. Xia (✉)
Key Lab of Theory and Technology for Advanced Batteries Materials, College of Engineering, Peking University, Beijing 100871, People's Republic of China
e-mail: dgxia@pku.edu.cn

B. Li
e-mail: libiao2008@pku.edu.cn

W. Huang · W. Chu · Z. Wu
National Synchrotron Radiation Laboratory, University of Science and Technology of China, Hefei 230029, People's Republic of China



B. Li, *Studies on Anionic Redox in Li-Rich Cathode Materials of Li-Ion Batteries*, Springer Theses, https://doi.org/10.1007/978-981-13-2847-3_3

have been made to realize improvement. Modification by doping other metal elements is effective to obtain good cyclability [12, 13]. However, such doping would usually cause a lowering of capacity owing to the incorporation of inactive elements. The coating approach can also improve rate capability and cycle stability for Li-rich layered oxides. But the highly continuous coating on the surface of particles is very difficult to obtain during practical process.

Herein, to realize the improvement of the electrochemical performance for $x Li_2MnO_3 \cdot (1 - x)Li[Ni_{1/3}Co_{1/3}Mn_{1/3}]O_2$, we put forward for the first time a strategy in modulating the electronic structure by the local incorporation of polyanions. The well-studied cathode material $0.5Li_2MnO_3 \cdot 0.5Li[Ni_{1/3}Co_{1/3}Mn_{1/3}]O_2$, which can also be written as $Li[Li_{0.2}Ni_{0.13}Co_{0.13}Mn_{0.54}]O_2$, was chosen as the target material to be modified since it can deliver a high capacity exceeding 300 mAh $g^{-1}$ but with poor stability. Polyanion-type compounds are a class of materials in which tetrahedral polyanion structure units $(XO_4)^{n-}$ and their derivatives $(X_mO_{3m+1})^{n-}$ (X = B, P, Si, S, As, Mo, or W) with strong covalent bonding combine with $MO_x$ (M = transition metal) polyhedral [14]. Their higher cyclability and thermal stability than conventional layered oxides, which is considered to be due to the high stability of oxygen ions, attract extensive attention, despite of their lower capacity [14]. Therefore, the merits of polyanion on stabilizing the oxygen close-packed structure inspired us to use nonmetal element, such as B, P, Si, S, to tune the electronic structure of the layered oxide. In this work, we present that, through incorporation of polyanions $(BO_3)^{3-}$ and $(BO_4)^{5-}$ in the Li-rich manganese-based layered material, the covalence of M–O decreases, and the top of O 2p band is lowered suggested by the first principle calculations, resulting in more stable oxygen ions under high voltage. The power of this approach is demonstrated by the facile synthesis of the Li-rich manganese-based oxide cathode material incorporated with boron polyanions, which shows much improved specific capacity, cycling performance, redox potentials, and thermal stability in comparison with pristine Li-rich cathode material.

## 3.2 Methods

### 3.2.1 Preparation of Bx-LRM (x = 0, 0.02, 0.04, 0.08)

$Li_{1.2}Ni_{0.13}Co_{0.13}Mn_{0.54}(BO_4)_{0.75x}(BO_3)_{0.25x}O_{2-3.75x}$ were synthesized via a sol–gel method to obtain the precursor. Stoichiometric amounts of lithium acetate dihydrate (5% excess), nickel acetate tetrahydrate, manganous acetate tetrahydrate, cobaltous acetate tetrahydrate, and boracic acid were mixed together and dissolved by deionized water with citric acid as the complexant and ethylene glycol as dispersant. The sol was evaporated using a rotary evaporator to form hydrous gel. The hydrous gel was then dried at 150 °C for 8 h. After milling the dried gel with a mortar and a pestle, it was sintered at 450 °C for 4 h and then at 850 °C for 15 h to get the target products.

### 3.2.2 Characterization of the Structures and Properties

To investigate the crystal structure of the layered materials before and after the substitution, X-ray diffraction (XRD) was taken using a D8 Advance diffractometer (Bruker, Germany), employing Cu Kα radiation ($\lambda$ = 1.5406 Å) operated at 40 kV and 40 mA. The scan data were collected in a $2\theta$ range of 10–80°, with a step size of 0.02° and a counting time of 4 s. Intermediate infrared spectra (4000−400 $cm^{-1}$) were recorded using Fourier transform infrared (FTIR) spectrometer (Bruker VECTOR22) equipped with MCT mid-IR, ATR, and so on. The powder was mixed with KBr to make a transparent slice for IR radiation, and the data were collected in absorption mode. The morphological characteristics of the powders were evaluated using cold field emission scanning electron microscopy (SEM, Hitachi S-4800) and transmission electron microscopy (TEM, FEI, TECNAI F20).

Hard XAS of Mn, Ni, Co K-edge spectra were collected in fluorescence mode at beamline 1W2B with a Si (111) double-crystal monochromator in Beijing Synchrotron Radiation Facility (BSRF). To prepare the samples during the charge process for ex situ measurement, all of the cells were charged and discharged at a current density of 60 mA $g^{-1}$ and then disassembled in the argon-filled glove box. The materials coated on an Al foil were sealed by 3 M tape before taken out and packed in a bag full of argon. Energy calibration was carried out by the first inflection point of Mn, Ni, and Co foil as the reference (Mn K-edge = 6539 eV, Co K-edge = 7709 eV and Ni K-edge = 8333 eV, respectively). The XAS data analysis was carried out using standard procedures by IFEFFIT. The spectrum below the pre-edge region was fitted to a straight line, and the post-edge background was subtracted based on the cubic spline procedure. The normalized EXAFS spectra were converted from E space to k space weighted by k and then converted to R space. Soft XAS data of Mn, Ni, Co L-edge, and O, B K-edge were collected in the total electron yield (TEY) mode at beamline 4B7B in Beijing Synchrotron Radiation Facility (BSRF). The energy calibration is performed by the $H_3BO_3$ standard sample. The peak-differentiating analysis of B K-edge XAS was performed by the software XPSPEAK41, with the value of sum of $\chi^2$ below 1. The data were analyzed first by background subtraction using the Shirley routine and a subsequent nonlinear fitting by mixed Gaussian–Lorentzian functions.

### 3.2.3 Electrochemical Tests

The cathode electrode was prepared by pasting the mixture of 80 wt% active material, 10wt% Super P, and 10wt% polyvinylidene fluoride (PVDF) on an aluminum foil. And then, the cathode electrode was dried at 120 °C overnight before the fabrication. The electrolyte was purchased from Beijing Institute of Chemical Reagents, the ingredient is confidential, and Whatman glass microfiber filters were used as the separator. The R2032-type coin cells were taken for the electrochemical testing and fabricated with the Li metal as the anode in the argon-filled glove box, in which the oxygen content and moisture content are all below 0.5 ppm. The galvanostatic charge/discharge process for all of the materials was measured by a Neware tester at the room temperature (25 °C), except for the high temperature (65 °C) tests. DSC measurements were performed to monitor the thermostability of these materials using a Mettler (DSC–1) instrument in a rate of 10 °C/min to 400 °C. Samples were prepared by charging the cells to 4.6 V at the second cycle and maintained at a constant voltage for 10 h, and then disassembled in an argon-filled glove box. The cathode material was sealed in an aluminum DSC pan with an additional 1 μL new electrolyte before taken out from glove box for measurement.

### *3.2.4 First Principle Calculations*

All the first principle calculations are performed with the Vienna ab inito simulation package (VASP) within the projector augmented wave (PAW) approach. The data are obtained within the GGA + U scheme. The Hubbard U value for Ni-3d, Co-3d, and Mn-3d is 6.30 eV, 4.91 eV, and 4.95 eV, respectively, according to the previous theoretical work. The spin polarization of ferromagnetic configuration is considered. The Monkhorst–Pack scheme 2 × 1 × 1 k-points mesh is used for the integration in the irreducible Brillouin zone. The energy cutoff for the plane waves is chosen to be 520 eV. The lattice parameters and the ionic position of the structure are fully relaxed, and the final forces on every relaxed atom are less than 0.05 eV $Å^{-1}$.

## 3.3 Results and Discussions

### *3.3.1 Structures and Properties of Bx-LRM (x = 0, 0.02, 0.04, 0.08)*

$Li[Li_{0.2}Ni_{0.13}Co_{0.13}Mn_{0.54}](BO_4)_{0.75x}(BO_3)_{0.25x}O_{2-3.75x}$ (B$x$-LRM, $x = 0, 0.02, 0.04$ and 0.08) were successfully synthesized via a conventional solid-state reaction at 850 °C. Figure 3.1 shows the X-ray diffraction (XRD) patterns of pristine and boracic polyanions-incorporated samples. All the peaks can be attributed to the layered α-$NaFeO_2$ structure of $R\overline{3}m$ space group with weak superlattice reflection peaks appearing at the 2θ range of 21–35°. No impurity peaks corresponding to the boron-related compounds were observed in the XRD patterns. With increasing amounts of the incorporated polyanions, both the superstructure-related peaks and main reflections corresponding to the layered structure become more and more intense, implying that more perfect layered structure and more ordering of Li–M–M in the transition metal layer were obtained. However, we did not find any prominent change of the lattice parameters even with the high boron content. This could be explained by the small radius of boron atom (0.27 Å) and its low scattering power due to its low atomic number, which will cause little change to the XRD pattern. On the other hand, the boron polyanion may exist as a local structure and will not affect the long-range order, as demonstrated later by XAS spectra. Figure 3.2 shows typical SEM and TEM images of the B$x$-LRM particles, which clearly indicate the size of the particles with smooth surface increasing from about 100 to 400 nm with $x = 0$, 0.02, 0.04, and 0.08 .

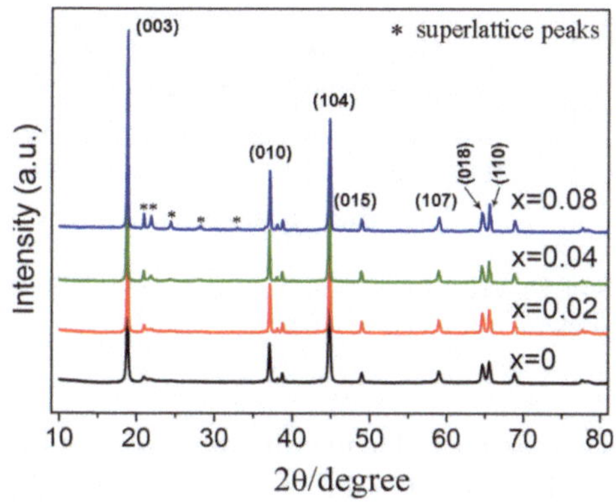


**Fig. 3.1** XRD patterns of B$x$-LRM ($x = 0$, 0.02, 0.04, 0.08)

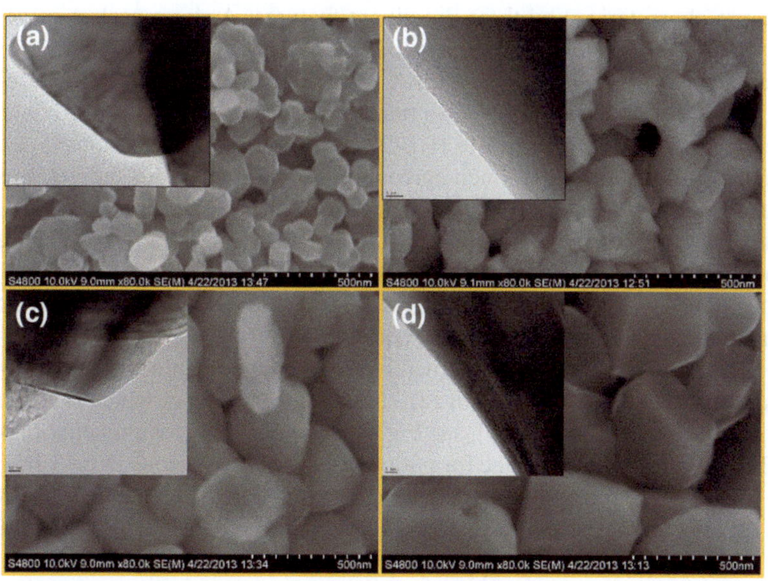


**Fig. 3.2** SEM and TEM figures of B$x$-LRM($x = 0$, 0.02, 0.04, 0.08)

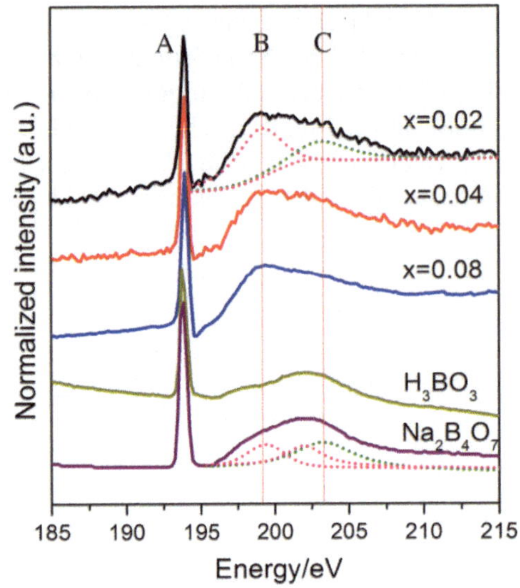


**Fig. 3.3** B K-edge XAS of B*x*-LRM for $x = 0$, 0.02, 0.04, and 0.08, with that of the standard samples $H_3BO_3$ and $Na_2B_4O_7$

B K-edge XANES was employed to identify the entries of trigonal and tetrahedral coordinated B species with oxygen in the as-prepared sample, as shown in Fig. 3.3. For comparison, B XAS spectra of the standard samples, $H_3BO_3$ and $Na_2B_4O_7$, were also measured. According to Fleet's work [15–17], the boron K-edge XANES spectra are characterized by three prominent features: A sharp peak A at ~194 eV corresponds to trigonal B ($^{III}B$); a broader peak B at 197–199 eV corresponds to tetrahedral B ($^{IV}B$); and a broad peak C contributed by $^{IV}B$ at 200–202 eV and $^{III}B$ at 203–204 eV. Figure 3.3 shows that $Na_2B_4O_7$ consists of both trigonal $BO_3$ and tetrahedral $BO_4$ groups, while $H_3BO_3$ with two significant features, A and C, comprises only $BO_3$ groups, which agrees well with the previous reports. Based on the analysis above, B*x*-LRM could be determined to contain both trigonal $BO_3$ and tetrahedral $BO_4$ groups. From the fitted dot line of different peaks (shown in Fig. 3.3), we quantified the ratio of $^{IV}B/^{III}B$ to be about 3 (3.05:1) in B*x*-LRM according to the peak area, since the ratio of $^{IV}B/^{III}B$ is 1:1 in $Na_2B_4O_7$. Hence, B*x*-LRM can be determined as $Li[Li_{0.2}Ni_{0.13}Co_{0.13}Mn_{0.54}](BO_4)_{0.75x}(BO_3)_{0.25x}O_{2-3.75x}$.

To further investigate the local structure around the polyanion, we performed the DFT simulations with a supercell of $Li_{45}[Li_9Ni_6Co_6Mn_{24}]O_{90}$, which is constructed based on previous report [18, 19]; see Fig. 3.4. Based on the assignment of $(BO_3)^{3-}$ and $(BO_4)^{5-}$ groups by XAS, the formation of boracic polyanions in B*x*-LRM could be the process of boron atoms bonded with the oxygen in the lattice of B0-LRM. $^{IV}B$ atoms may occupy the tetrahedral interstitial sites formed by three planes of adjacent $MO_6$ octahedrons and one plane of a $LiO_6$ octahedron, while $^{III}B$ has many kinds of

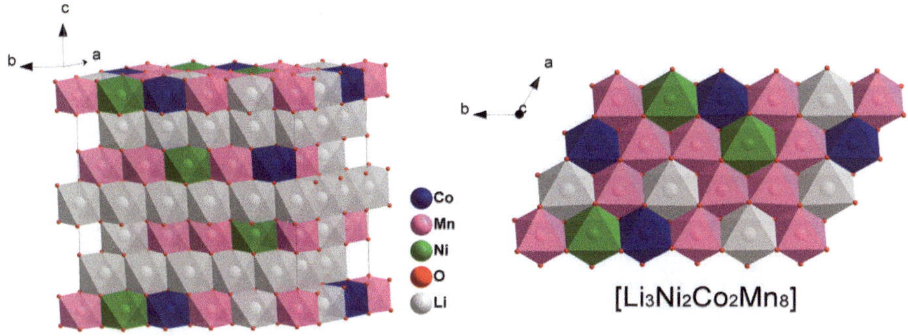


**Fig. 3.4** Supercell of $Li_{45}[Li_9Ni_6Co_6Mn_{24}]O_{90}$ and its atomic arrangement in transition metal layer

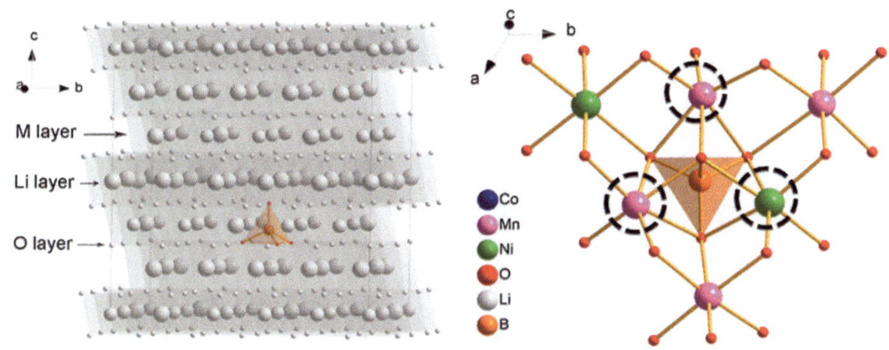


**Fig. 3.5** Initial structure of Li45[Li9Ni6Co6Mn24]BO90 for optimization

possible sites; such as one consists of three equivalent apical oxygen atoms of three adjacent $MO_6$ octahedrons, and another is the central site formed by a triangle plane of a $MO_6$ plane. Furthermore, $^{IV}B$ is the dominant species, so we chose the structure with a $^{IV}B$ to start the optimization, as shown in Fig. 3.5. The calculation will bring the optimized and the most stable structure to us.

We optimized four different boron tetrahedral sites in total and found that three of them remained as $^{IV}B$ while the other one relaxed to be $^{III}B$; see Fig. 3.6. Note that this proportion of 3:1 is consistent with the ratio quantified by boron XANES spectra. The simulated local structures around the boracic polyanion show that the M–O bonds are lengthened due to the strong B–O bond, inducing the local octahedron geometrical distortion. The variation of the electronic structure, caused by the structure distortion simultaneously, will be discussed later.

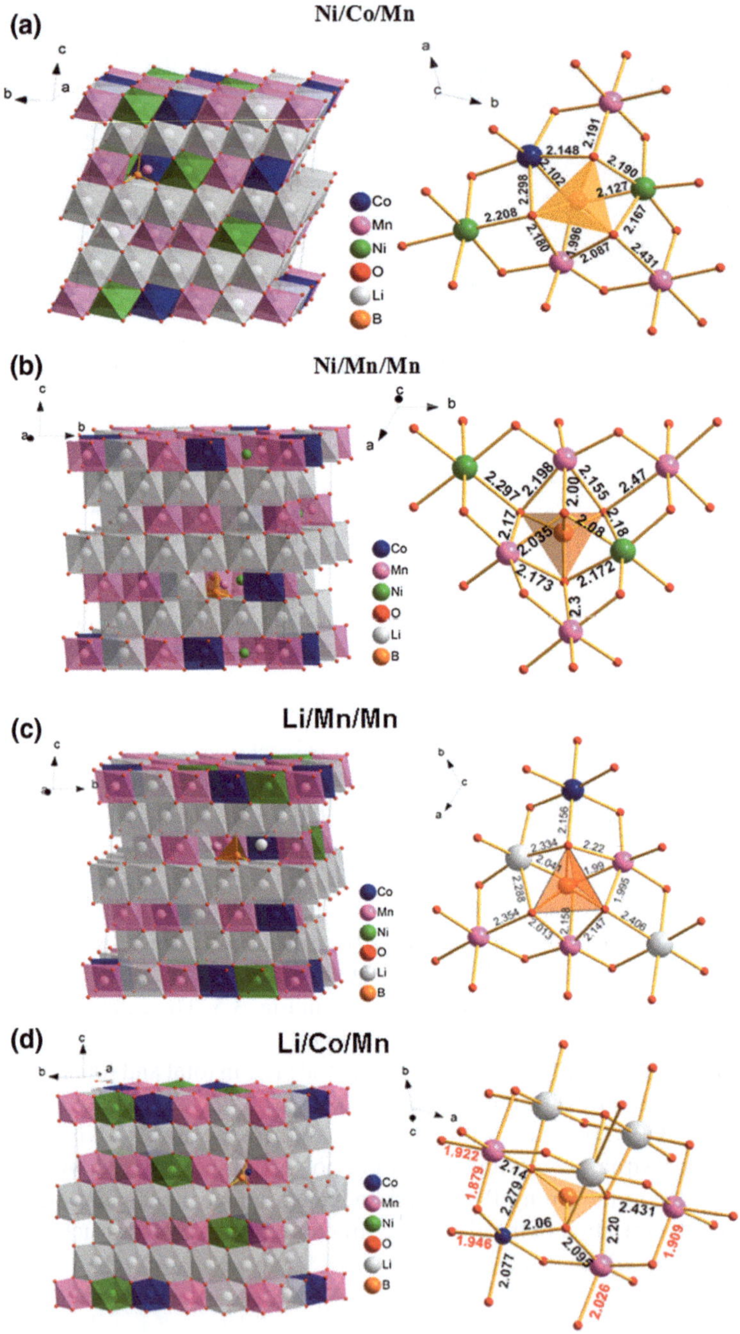


**Fig. 3.6** Optimized structure of $Li_{45}[Li_9Ni_6Co_6Mn_{24}]BO_{90}$ with different tetrahedral boron sites. The local structure consists of three transitional metal atoms of **a** Ni/Co/Mn; **b** Ni/Mn/Mn; **c** Li/Mn/Mn; **d** Li/Co/Mn

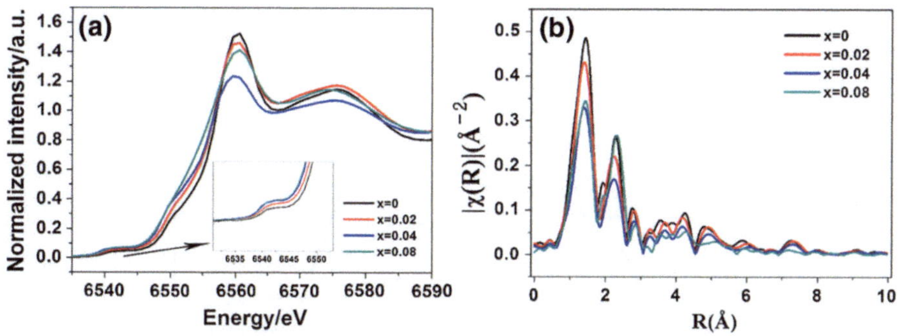


**Fig. 3.7** **a** Mn K-edge XANES of B$x$-LRM with $x = 0$, 0.02, 0.04, and 0.08; **b** Fourier transform (FT) magnitudes of B$x$-LRM with $x = 0$, 0.02, 0.04, and 0.08

K-edge XAFS spectra of Mn, Ni, and Co were measured to further reveal the local environment changes of transition metals after boracic polyanion's incorporation. Figure 3.7 shows that the adsorption edge shifts to lower energy with$x$ increases for Mn K-edge XANES spectrum, suggesting that the boracic polyanion incorporation brings the valence of Mn to decrease in order to rebuild the charge balance of the system. Moreover, the decreased intensity of the white line peak indicates a large variation of the $MnO_6$ octahedral environment [20]. In addition, the pre-edge peak also exhibits an increase of intensity as a function of$x$, as clearly shown in the inset of Fig. 3.7a. The pre-edge peak is corresponding to the electronic transition from 1 s core levels to the 4p components of 3d–4p hybridized states when the $MnO_6$ octahedral distortion appears, and more distorted the octahedron is, stronger the peak intensity of the pre-edge exhibits [21, 22]. Therefore, the increased intensity of the pre-edge peak with high$x$ value indicates a more distorted $MnO_6$ octahedron due to the boron's strong binding force with oxygen. The Fourier-transformed radial structure functions of Mn K-edge XAFS (Fig. 3.7b) and FTIR absorption spectra (Fig. 3.8) also confirm the intensive local disorder of $MnO_6$ structure and the presence of distorted $MO_6$ octahedron [23], respectively, after boron's incorporation. However, different from the case of Mn, the K-edge XAFS spectra of Ni and Co (Fig. 3.9) display little change, which may be due to their lower concentrations compared with Mn and the consequent lower probability to be affected by boron.

The electrochemical measurements on the Li-ion cells were performed by galvanostatic charge/discharge technique. Figure 3.10a compares the cyclability of B0-LRM and B0.02-LRM ($x = 0.02$), under a 0.1 C (20 mA/g) current density in the voltage range of 2.0–4.8 V. The former presents a poor cycling performance with only 70% capacity retention, delivering 298 mAh $g^{-1}$ at the first cycle and only

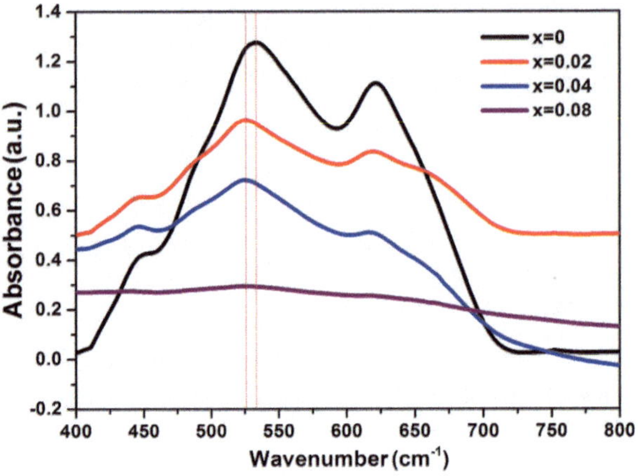


**Fig. 3.8** FTIR spectra of B$x$-LRM with $x$ = 0, 0.02, 0.04, and 0.08

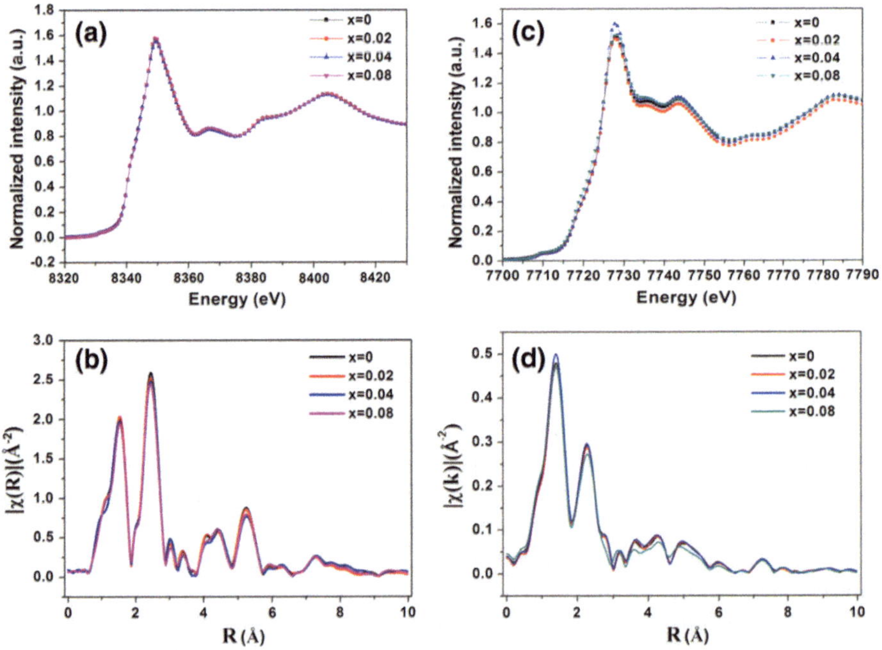


**Fig. 3.9** **a** Ni K-edge XANES of B$x$-LRM with $x$ = 0, 0.02, 0.04, and 0.08; **b** Ni Fourier transform (FT) magnitudes of B$x$-LRM with $x$ = 0, 0.02, 0.04, and 0.08; **c** Co K-edge XANES of B$x$-LRM with $x$ = 0, 0.02, 0.04, and 0.08; **d** Co Fourier transform (FT) magnitudes of B$x$-LRM with $x$ = 0, 0.02, 0.04, and 0.08

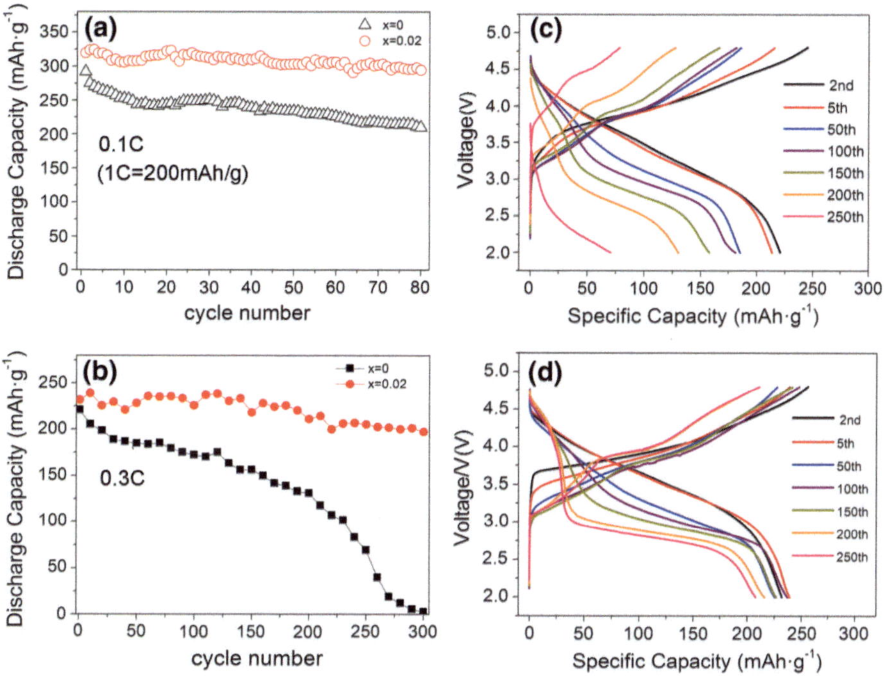


**Fig. 3.10** Electrochemical performance of B$x$-LRM. Cycling performance of B$x$-LRM with $x$ = 0 and 0.02 at a current density of **a** 20 mA $g^{-1}$ and **b** 60 mA $g^{-1}$; Voltage evolution of B$x$-LRM with **c** $x = 0$ and **d** $x = 0.02$ at a current density of 60 mA $g^{-1}$

210 mAh $g^{-1}$ at the 80th cycle. The polyanion-modified material exhibits a superior cycling performance of 94% capacity retention, with 319 mAh $g^{-1}$ at the first cycle and about 300 mAh $g^{-1}$ at the 80th cycle. With the increase of $x$, the capacity is reduced to 250 mAh $g^{-1}$ for $x = 0.04$ and 200 mAh $g^{-1}$ for $x = 0.08$ at the first cycle, but recovers to about 250 mAh $g^{-1}$ at the 80th cycle (Fig. 3.11a). The lower capacity of the initial cycles when $x = 0.08$ may be attributed to the insufficient activation of the $Li_2MnO_3$ component. But during the subsequent cycles, the activation process is continued, with the capacity increases concurrently. Figure 3.10b further compares the long cycling performance of B0-LRM and B0.02-LRM within 300 cycles (that of B0.04-LRM and B0.08-LRM are shown in Fig. 3.11b). The B0-LRM has large capacity fading especially after 200 cycles, while for the B0.02-LRM, a capacity retention of about 89% was obtained, with a reversible capacity of 200 mAh $g^{-1}$ within 300 cycles under a 0.3 C current density of 60 mA $g^{-1}$. However, the rate performance of B0.02-LRM is not much improved since the particle grew larger than that of B0-LRM; see Fig. 3.11c and d.

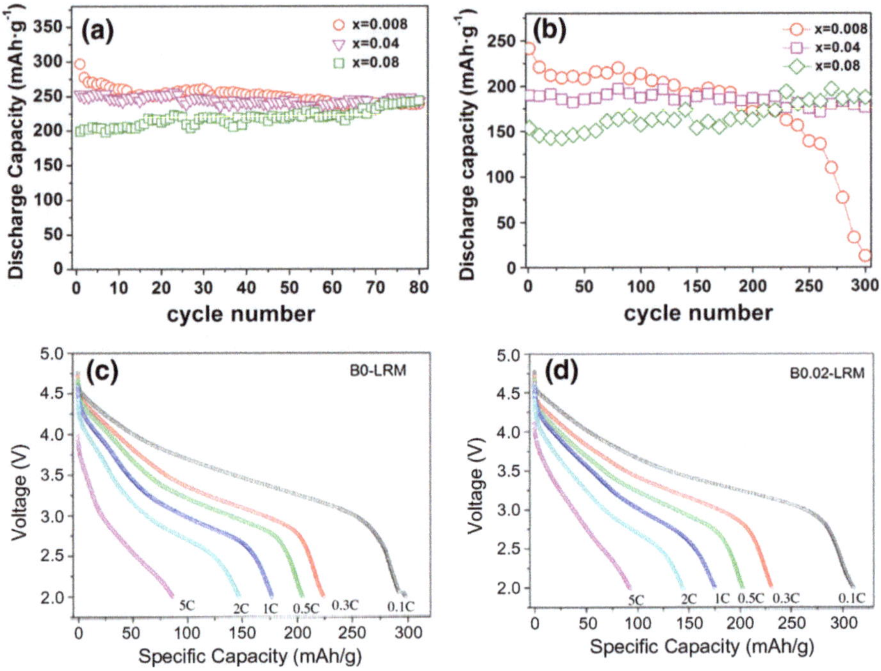


**Fig. 3.11** Cycling performance of B$x$-LRM with $x = 0.008$, 0.04, and 0.08 at a current density of **a** 20 mA/g and **b** 60 mA/g in the voltage range of 2.0–4.8 V. Rate performance of **c** B0-LRM and **d** B0.02-LRM

Moreover, the voltage degradation of B$x$-LRM is also largely buffered, as can be seen from the comparison of Fig. 3.10c and d. After 300 cycles under a current density of 60 mA $g^{-1}$, the B0.02-LRM still retains an average discharge voltage of 2.7 V, whereas the B0-LRM only shows a drastically voltage linear decline. Figure 3.12 plots the dQ/dV profiles of the fifth cycle for B$x$-LRM. It can be found that the redox couple potentials are enhanced substantially after the incorporation. For the comparison between B0-LRM and B0.02-LRM, the redox couples of $Mn^{4+}/Mn^{3+}$ and $Ni^{4+}/Ni^{3+}/Ni^{2+}$ increase from 3.02 and 3.61 V to 3.44 and 3.76 V for the discharge process, respectively, and $Co^{4+}/Co^{3+}$ also shows slight enhancement. This leads to a higher energy density of B0.02-LRM than B0-LRM.

The Differential Scanning Calorimetry (DSC) measurement was performed to reveal the material's thermal stability, which is one of the indicators that determine the safety characteristics of the battery at elevated temperatures. Figure 3.13 shows that the electrodes of B0.02-LRM, charged to 4.6 V, possess higher exothermic reaction temperature than that of B0-LRM, presenting intrinsically safe characteristics after the incorporation of boracic polyanion. Furthermore, we also tested the cells at 65 °C to investigate the cycling performance, and the sample B0.02-LRM has higher stability than B0-LRM; see in Fig. 3.14.

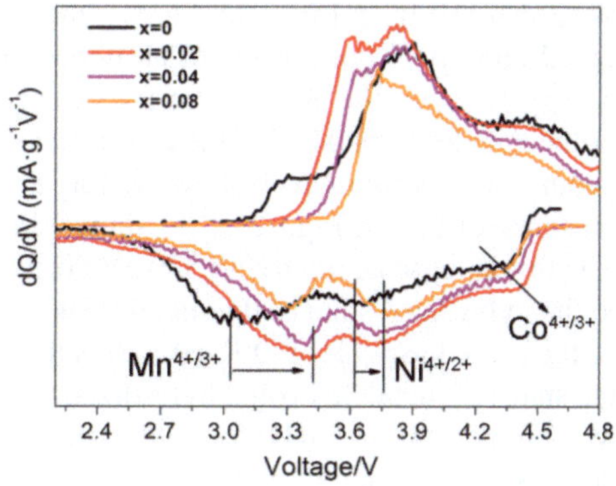


**Fig. 3.12** dQ/dV profiles of the fifth cycle of B*x*-LRM at a current density of 60 mA $g^{-1}$

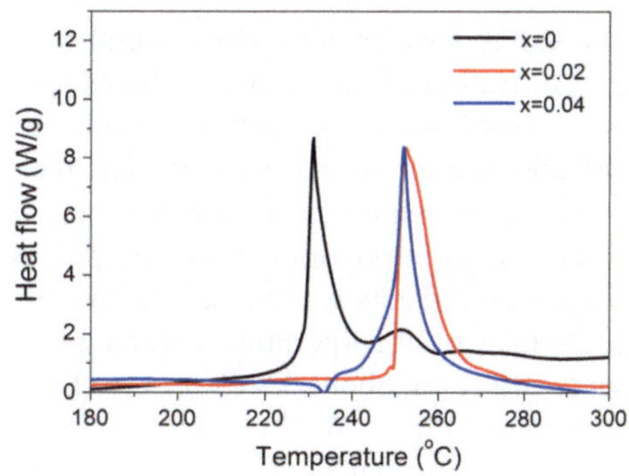


**Fig. 3.13** DSC profiles of B*x*-LRM charged to 4.6 V at the second cycle and kept for 10 h with constant voltage at a scanning rate of 10 °C/min

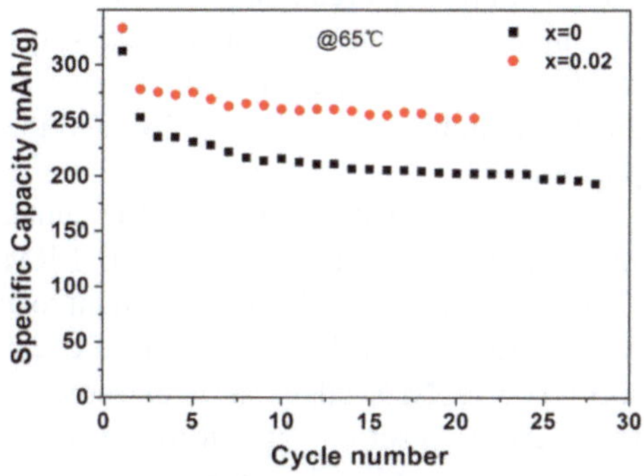


**Fig. 3.14** Cycling performance of B0-LRM and B0.02-LRM at 65 °C

The much improved electrochemical performance of the B0.02-LRM in comparison with B0-LRM could be attributed to the electronic structural changes resulted from the proposed structural design. Figure 3.15 shows the soft XAS spectra of Ni, Co, Mn, and O of B*x*-LRM, respectively. It can be found that the $L_{II}$, $L_{III}$ edges of Mn, Co, and Ni for B*x*-LRM show an evident shift to lower energy as *x* increases, as shown in Fig. 3.15a, b and c. $L_{II}$ and $L_{III}$ could be assigned to the transition of 2p3/2 and 2p1/2 core levels to an unoccupied 3d hole [20, 24]. The shift to lower energy indicates the decrease of the valence of the transition metal and the decrease of the covalence of M–O. This electronic structural change corresponds to the local geometrical variation well shown before as the M–O bonds are lengthened due to the presence of boron. Figure 3.15d shows the K-edge spectra of O in these materials. The first intense peak around 530 eV corresponds to the transition of O 1 s electrons to the hybridized state of the metal 3d with O 2p orbitals [25]. The integral intensity of the pre-edge peak of O K-edge spectra depends on the total d electronic holes in the states of metal 3d orbital hybridized with the O 2p orbital, or the covalency of the M–O bond [24]. It can be found that the integral intensity of the pre-edge peak of O decreases dramatically with respect to the$x$ value. Combined with the metal L-edge results, this intensity change further confirms the decreased covalency of M–O bonds. The strong covalency of active metal with oxygen, especially for Co–O, facilitates the generation of the electronic holes for the 2p band of oxygen and results in the evolution of oxygen under high delithiated state. Here, the decreased M–O covalence will alleviate the variation of oxygen 2p band during cycling, which can be assigned to one of the main reasons for the improved stability. Furthermore, as proposed by Goodenough in previous work, the polyanions $(X_mO_{3m+1})^{n-}$ will lower the top of O 2p band effectively compared with that in layered oxides [3, 14, 26]. We demonstrate that, by incorporating boracic polyanions, the top of O 2p band is lowered successfully for about 0.8 eV in $Li_{45}[Li_9Ni_6Co_6Mn_{24}]BO_{90}$ either with $^{IV}B$ or $^{III}B$, as shown in Fig. 3.16. Though the top of O 2p band is still largely higher than totally polyanion-type materials like $LiMPO_4$, this will remarkably enhance the oxygen ion stability in this layered material under high voltage. As to the enhancement of redox potential, it could be attributed to the strong polarization of oxygen atoms to boron, resulting in the energies of antibonding d-electron redox couples of M being lowered due to the inductive effect [14, 26], and the stronger the polarization, the higher the potential.

To further reveal the origin of the enhanced stability, Ni, Co, and Mn K-edge ex situ XANES of B0-LRM and B0.02-LRM were collected, respectively. We chose the tenth cycle to take the measurement to avoid the interference caused by the structure rearrangement during the initial cycles; see Fig. 3.17. We can find that in the second cycle, Ni K-edge has an extremely abnormal behavior of shifting back to the lower energy position, similar to the case of first cycle, implying that the structural rearrangement still exists, while for the tenth cycle, the K-edge of Ni shifts gradually for both of B0-LRM and B0.02-LRM, with no back-shift behavior appearing anymore. This indicates that in the tenth cycle, the structural variation induced by the activation of $Li_2MnO_3$ can be ignored. So our analysis of the comparison of the structural changes between B0-LRM and B0.02-LRM based on the tenth cycle is

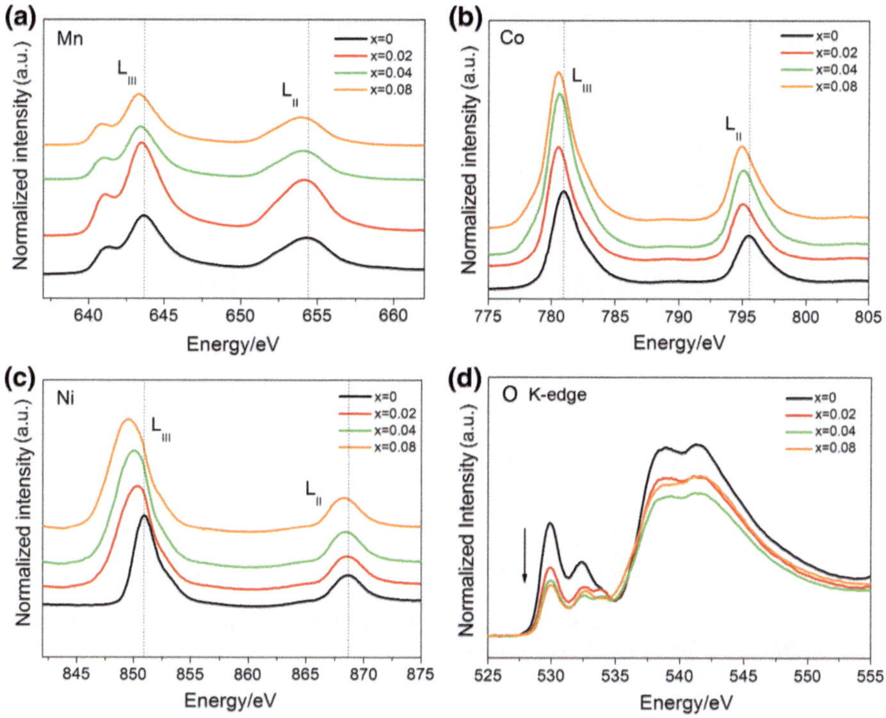


**Fig. 3.15** Electronic structural variations of B$x$-LRM. L-edge XANES spectra of **a** Mn, **b** Co, **c** Ni, and **d** O K-edge spectra in B$x$-LRM, $x = 0$, 0.02, 0.04, and 0.08

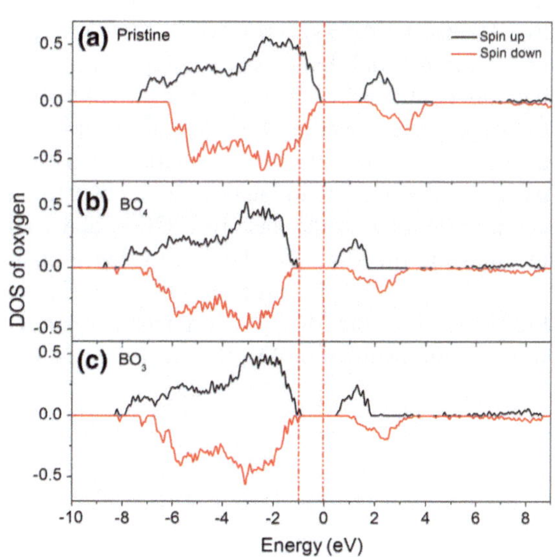


**Fig. 3.16** Calculated O DOS of B$x$-LRM. Calculated O DOS of **a** $Li_{45}[Li_9Ni_6Co_6Mn_{24}]O_{90}$, **b** $Li_{45}[Li_9Ni_6Co_6Mn_{24}](BO_4)O_{86}$ with $^{IV}B$ in the site of "Li/Mn/Mn", and **c** $Li_{45}[Li_9Ni_6Co_6Mn_{24}](BO_3)O_{87}$ with $^{III}B$ in the site of "Li/Co/Mn"

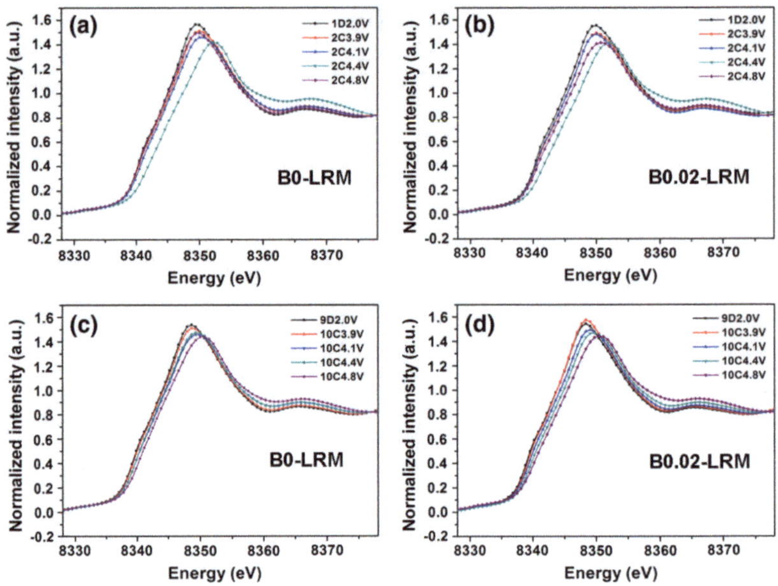


**Fig. 3.17** Ni K-edge XAFS of **a** B0-LRM and **b** B0.02-LRM in the second cycle; Ni K-edge XAFS of **c** B0-LRM and **d** B0.02-LRM in the tenth cycle

persuasive, since we avoid the interference of the activation of $Li_2MnO_3$. Co K-edge spectra in B0.02-LRM show a larger shift of ~1 eV than the case in pristine B0-LRM (Fig. 3.18a). Furthermore, the first peak in the radial structure functions of Fourier-transformed EXAFS spectrum of B0.02-LRM (Fig. 3.18d) demonstrates less change under different delithiated states compared with that of B0-LRM (Fig. 3.18c), implying a more stable Co–O octahedron environment. For B0-LRM, it is suggested that the large energy overlap of $Co^{4+}/Co^{3+}$ couple with O 2p band makes Co take little change while most of the contribution of the charge compensation is realized by oxygen upon high delithiated state, leading to the oxygen evolution and structural collapse, similar to the case in $LiCoO_2$ and $LiNi_{1/3}Co_{1/3}Mn_{1/3}O_2$. However, after the substitution, Co will take large valence change to compensate for the charge, and little holes will be generated on the O 2p band; see the schematic diagram in Fig. 3.19. Thus, the $MO_6$ environment and layered structure are stabilized, which is further demonstrated by Mn K-edge ex situ XAS, displayed in Fig. 3.20.

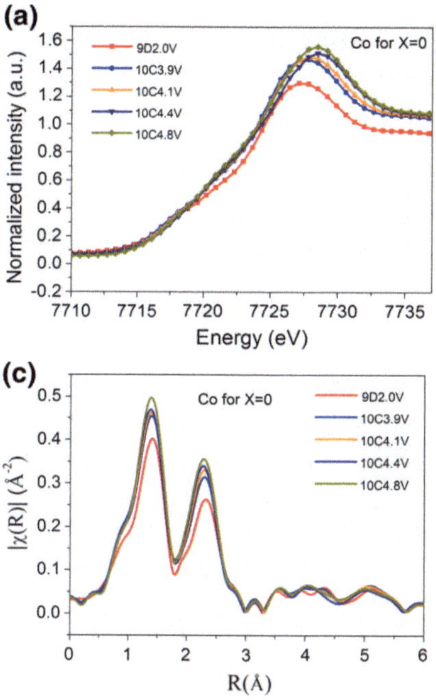


**Fig. 3.18** Ex situ XAS of B$x$-LRM during the tenth charge process. Co K-edge XANES of **a** B0-LRM and **c** B0.02-LRM; **b** and **d** are their Fourier-transformed EXAFS results during the tenth charge process, respectively

**Fig. 3.19** Schematic diagram of the charge compensation mechanism before and after the boracic polyanions substitution

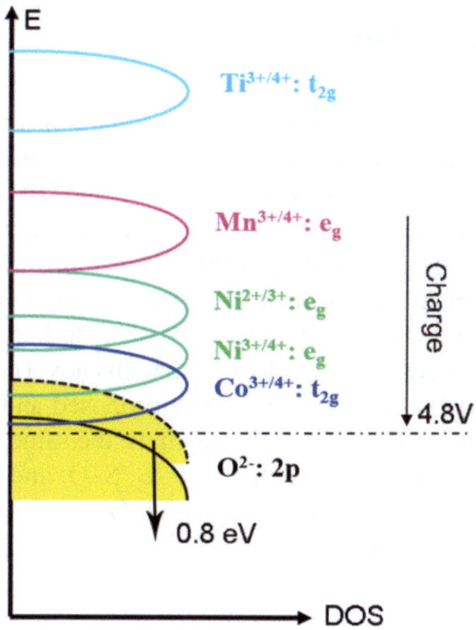

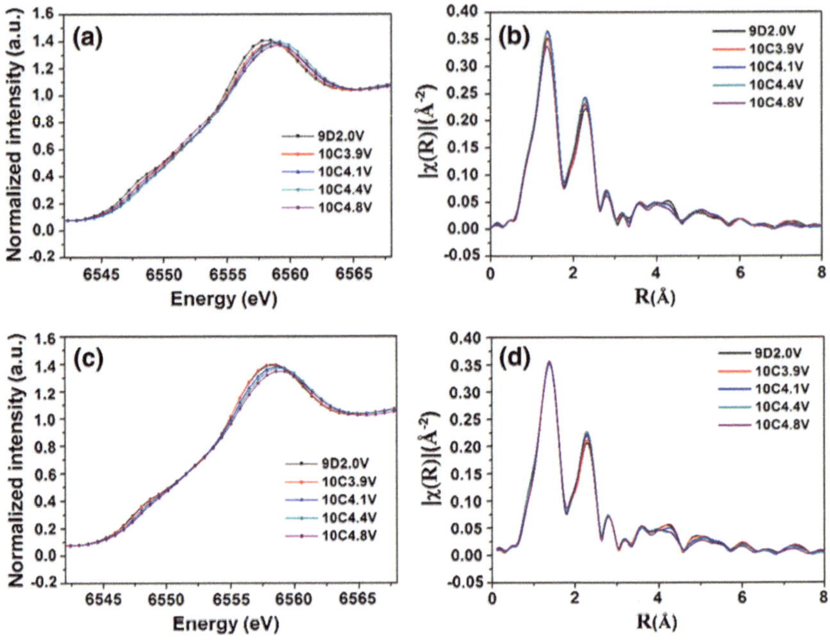


**Fig. 3.20** **a** Mn K-edge XANES of LRM and its **b** Fourier-transformed EXAFS result; **c** Mn K-edge XANES of B0.02-LRM and its **d** Fourier-transformed EXAFS result during the tenth charge process

## 3.4 Summary

In summary, we successfully demonstrated a new strategy to improve the performance of the layered cathode by manipulating the electronic structure through incorporation of polyanion $(BO_3)^{3-}$ and $(BO_4)^{5-}$. The materials showed much improved specific capacity, cycling performance, enhanced redox potentials, and superior thermal stability. We attributed this improved performance to the decreased covalency of the M–O bonds and the lowered top of O 2p band, which may alleviate the variation on the electronic structure of O 2p band induced during delithiation and lithiation process. We should note that this novel combination of polyanion with layered materials is the first time to be proposed and realized. Furthermore, the polyanions like $PO_4$, $SO_4$, and $SiO_4$ may have the same possibility to realize this enhancement, and the strong covalency of P–O, S–O, and Si–O may have the same effect to tune the electronic structure of the M–O systems, which will be extended to other work about cathode preparation.

# References


1. Kang K, Meng YS, Breger J, et al. Electrodes with high power and high capacity for rechargeable lithium batteries. Science. 2006;311(5763):977–80.
2. Sun YK, Myung ST, Park BC, et al. High-energy cathode material for long-life and safe lithium batteries. Nat Mater. 2009;8(4):320–4.
3. Goodenough JB, Park KS. The Li-ion rechargeable battery: a perspective. J Am Chem Soc. 2013;135(4):1167–76.
4. Shen C-H, Wang Q, Chen H-J, et al. In situ multitechnical investigation into capacity fading of high-voltage $LiNi_{0.5}Co_{0.2}Mn_{0.3}O_2$. ACS Appl Mater Interfaces. 2016;8(51):35323–35.
5. Wei GZ, Lu X, Ke FS, et al. Crystal habit-tuned nanoplate material of $Li[Li_{1/3-2x/3}Ni_xMn_{2/3-x/3}]O_2$ for high-rate performance lithium-ion batteries. Adv Mater. 2010;22(39):4364–7.
6. Yabuuchi N, Yoshii K, Myung ST, et al. Detailed studies of a high-capacity electrode material for rechargeable batteries, $Li_2MnO_3$-$LiCo_{1/3}Ni_{1/3}Mn_{1/3}O_2$. J Am Chem Soc. 2011;133(12):4404–19.
7. Lee Y, Kim MG, Cho J. Layered $Li_{0.88}[Li_{0.18}Co_{0.33}Mn_{0.49}]O_2$ nanowires for fast and high capacity Li-ion storage material. Nano Lett. 2008;8(3):957–61.
8. Kim MG, Jo M, Hong YS, et al. Template-free synthesis of $Li[Ni_{0.25}Li_{0.15}Mn_{0.6}]O_2$ nanowires for high performance lithium battery cathode. Chem Commun. 2009;2:218–20.
9. Wang D, Belharouak I, Gallagher S, et al. Chemistry and electrochemistry of concentric ring cathode $Li_{1.42}Ni_{0.25}Mn_{0.75}O_{2+\gamma}$ for lithium batteries. J Mater Chem. 2012;22(24):12039.
10. Gu M, Belharouak I, Zheng J, et al. Formation of the spinel phase in the layered composite cathode used in Li-ion batteries. ACS Nano. 2013;7(1):760–7.
11. Zheng J, Gu M, Xiao J, et al. Corrosion/fragmentation of layered composite cathode and related capacity/voltage fading during cycling process. Nano Lett. 2013;13(8):3824–30.
12. Sun YK, Lee MJ, Yoon CS, et al. The role of $AlF_3$ coatings in improving electrochemical cycling of Li-enriched nickel-manganese oxide electrodes for Li-ion batteries. Adv Mater. 2012;24(9):1192–6.
13. Wu F, Li N, Su Y, et al. Spinel/layered heterostructured cathode material for high-capacity and high-rate Li-ion batteries. Adv Mater. 2013;25(27):3722–6.
14. Padhi AK, Nanjundaswamy KS, Goodenough JB. Phospho-olivines as positive-electrode materials for rechargeable lithium batteries. J Electrochem Soc. 1997;144(4):1188–94.
15. Li D, Bancroft GM, Fleet ME. B K-edge XANES of crystalline and amorphous inorganic materials. J Electron Spectros Relat Phenom. 1996;79:71–3.
16. Fleet ME, Muthupari S. Coordination of boron in alkali borosilicate glasses using XANES. J Non-Cryst Solids. 1999;255(2–3):233–41.
17. Fleet Michael E, Muthupari S American mineralogist, vol. 85. 2000. p. 1009.
18. Jarvis KA, Deng Z, Allard LF, et al. Atomic structure of a lithium-rich layered oxide material for lithium-ion batteries: evidence of a solid solution. Chem Mater. 2011;23(16):3614–21.
19. Boulineau A, Simonin L, Colin JF, et al. First evidence of manganese-nickel segregation and densification upon cycling in Li-rich layered oxides for lithium batteries. Nano Lett. 2013;13(8):3857–63.
20. Simmance K, Sankar G, Bell RG, et al. Tracking the formation of cobalt substituted $AlPO_4$ using simultaneous in situ X-ray diffraction and X-ray absorption spectroscopy techniques. Phys Chem Chem Phys. 2010;12(3):559–62.
21. Yamamoto T. Assignment of pre-edge peaks in K-edge X-ray absorption spectra of 3d transition metal compounds: electric dipole or quadrupole. X-Ray Spectrom. 2008;37(6):572–84.
22. Wu ZY, Xian DC, Hu TD, et al. Quadrupolar transitions and medium-range-order effects in metalk-edge X-ray absorption spectra of 3d transition-metal compounds. Phys Rev B. 2004;70(3).

23. Qiao QQ, Zhang HZ, Li GR, et al. Surface modification of Li-rich layered $Li(Li_{0.17}Ni_{0.25}Mn_{0.58})O_2$ oxide with $LiMnPO_4$ as the cathode for lithium-ion batteries. J Mater Chem A. 2013;1(17):5262.
24. Ito A, Sato Y, Sanada T, et al. In situ X-ray absorption spectroscopic study of Li-rich layered cathode material $Li[Ni_{0.17}Li_{0.2}Co_{0.07}Mn_{0.56}]O_2$. J Power Sources. 2011;196(16):6828–34.
25. Hy S, Su W-N, Chen J-M, et al. Soft X-ray absorption spectroscopic and raman studies on $Li_{1.2}Ni_{0.2}Mn_{0.6}O_2$ for lithium-ion batteries. J Phys Chem C. 2012;116(48):25242–7.
26. Padhi AK, Nanjundaswamy KS, Masquelier C, et al. Effect of structure on the $Fe^{3+}/Fe^{2+}$ redox couple in iron phosphates. J Electrochem Soc. 1997;144(5):1609–13.

# Chapter 4
# Anionic Redox and Stability Mechanism of Li-Rich Layered Oxides

**Biao Li, Ruiwen Shao, Huijun Yan, Li An, Bin Zhang, Hang Wei, Jin Ma, Dingguo Xia and Xiaodong Han**

## 4.1 Introduction

Lithium-ion batteries have been increasingly urged with high-energy density and long cycle life to meet the increasing requirements for portable electronics, use of renewable energy, and electric vehicles [1, 2]. The availability of the appropriate cathode material is crucial for the development of lithium-ion batteries with high-energy density. Although the conventional layered cathode such as $LiCoO_2$, $LiNiO_2$, and $LiNi_{1/3}Co_{1/3}Mn_{1/3}O_2$ has a theoretical capacity of more than 250 mA h $g^{-1}$, they cannot retain their layered structure after deep charging/discharging, owing to the excessive participation of oxygen redox. To obtain high-capacity cathode with good cycles, numerous researchers have emphasized the importance of developing cathode materials with well-ordered structures where most of doping metal atoms exist in the transition metal (TM) sublattice. However, the conventional layered oxide cathode after modification can only deliver a practical capacity of less than 200 mA h $g^{-1}$. In recent years, lithium-rich layered oxides based on $Li_2MO_3$ (M = Mn, Ru, etc.) with layered structure are considered as promising cathodes for Li-ion batteries due to

This chapter was adapted from B. Li et al. *Advanced Functional Materials* 2016, 26 (9), 1330–1337 with permission. Copyright 2016, Wiley-VCH.

B. Li · H. Yan · L. An · H. Wei · J. Ma · D. Xia (✉)
Beijing Key Lab of Theory and Technology for Advanced Batteries Materials, College of Engineering, Peking University, Beijing 100871, P. R. China
e-mail: dgxia@pku.edu.cn

B. Li
e-mail: libiao2008@pku.edu.cn

R. Shao · B. Zhang · X. Han
Institute of Microstructure and Properties of Advanced Materials, Beijing University of Technology, Beijing 100124, P. R. China



B. Li, *Studies on Anionic Redox in Li-Rich Cathode Materials of Li-Ion Batteries*,
Springer Theses, https://doi.org/10.1007/978-981-13-2847-3_4

their higher capacity of more than 220 mA h $g^{-1}$ and higher stability at high voltage as compared with the conventional $LiMO_2$ layered oxides, provoking worldwide attentions [3–9]. The common feature accounting for the high capacities of these lithium-rich layered oxides is the two redox processes involving both of cation and anion [10–13]. Despite the same process of partly charge compensation by oxygen during deep charging/discharging process, the lithium-rich layered oxides can deliver much higher reversible capacity than that of conventional cathode materials [14–24]. To the best of our knowledge, such difference on the structural stability between the conventional and lithium-rich layered oxides has been attracted limited attention. It is of high importance that there must be some unknown mechanism leading to the stability difference since the geometric and electronic structure between these two kinds of layered materials is similar, except for the superlattice structure. In present work, we elucidate why lithium-rich layered oxides $Li_2RuO_3$ exhibit high capacities without undergoing a structural collapse for a certain number of cycles by tracking the electronic and geometric structural changes induced in $Li_2RuO_3$ during charging/discharging. Backed up the density functional theory calculations and a series of in situ experiments, we unravel that the presence of the lithium atoms in the transition metal layer could make the lithium-rich structure flexible, facilitating the formation of $O_2^{2-}$-like species, which favors the structural integrity and providing high capacity.

## 4.2 Methods

### 4.2.1 *Preparation of $Li_2RuO_3$*

Appropriate amounts of $Li_2CO_3$ (Sinopharm, 99.99%) and $RuO_2{\bullet}xH_2O$ (J&K, 54% Ru) were ground and mixed homogeneously after being heated at 300 °C for 4 h to remove the crystal water. The product, $Li_2RuO_3$, was obtained by heating the mixture at 900 °C for 12 h and then at 1000 °C for 12 h in air after being cooled.

### 4.2.2 *Characterization of the Structures and Properties*

The morphology of the $Li_2RuO_3$ particles was characterized using cold field emission scanning electron microscopy (SEM) (Hitachi S-4800) and TEM (FEI, TECNAI F20). The XRD spectrum of the $Li_2RuO_3$ particles was measured using a D8 Advance diffractometer (Bruker, Germany) with a Cu Kα radiation ($\lambda$ = 1.5406 Å) source; the system was operated at 40 kV and 40 mA. The data were collected for 2θ values of 10–80°; the step size was 0.02°; and the counting time was 0.5 s.

### 4.2.3 *Preparation of the Electrodes and Electrochemical Tests*

Cathode electrodes were prepared by mixing 80% $Li_2RuO_3$ powder, 10% Super P, and 10% polytetrafluoroethylene (used as the binder) homogeneously and rolling the mixture into a slice before cutting it into pieces. R2032-type coin cells were used for the galvanostatic charge/discharge tests; a lithium metal plate was employed as the anode and Whatman GF/D glass microfiber filters as the separator. The electrolyte used was purchased from the Beijing Institute of Chemical Reagents; its ingredients are confidential. All the cells, including the in situ cells, were fabricated in a Braun argon-filled glove box; the moisture and oxygen levels were lower than 0.1 ppm. The galvanostatic charge/discharge measurements were performed at room temperature (25 °C) using a Neware tester.

### 4.2.4 *In Situ Experimental Methods*

The in situ XRD patterns were measured by fabricating $Li_2RuO_3$ electrodes; lithium metal was used as the anode. A laboratory-designed stainless-steel cell with a beryllium window as the current collector on the cathode side, allowing the X-rays to penetrate the cell, was used. The in situ XRD patterns were collected as the cell was being charged and discharged slowly at a rate of 0.1 °C, in order to capture the static or quasi-static evolutions. 95 XRD patterns were obtained from the open circuit voltage once the cell was fabricated to the end of the second charge process.

The in situ Ru K-edge XAS spectra were collected in the transmission mode at beamline BL14W1 of the Shanghai Synchrotron Radiation Facility and at beamline 1W1B of the Beijing Synchrotron Radiation Facility, China, using a Si (311) double-crystal monochromator. The in situ cells were designed in the laboratory on the basis of the cell used for the in situ XRD measurements; a Be window was formed on both the cathode side and the anode side to allow the X-ray beam to penetrate. The spectrum was recorded per 15 min simultaneously with the cell charged/discharged at a low current density of 30 mA $g^{-1}$; 98 spectra were acquired from the start of the first charge process to the end of the second charge process. All data processing performed prior to analysis, including energy calibration, background removal, normalization, and Fourier transformation, was done using the software Athena. The first-shell extended X-ray absorption fine structure (EXAFS) fittings were made using the program Artemis.

In situ TEM and EELS were performed using a JEOL-2010 TEM system with an accelerating voltage of 200 kV and a commercial scanning tunneling microscopy-TEM system (Nanofactory Instruments). To fabricate the test batteries, $Li_2RuO_3$ particles were made to adhere to the tip of a gold pin and were used as the cathode, while Li metal was attached to a tungsten rod and used as the anode; $Li_2O$ coated on the surface of Li was used as the solid electrolyte. The internal circuit was completed by making the gold rod dipped in Li touch the selected particles of $Li_2RuO_3$ on the tungsten tip. The in situ test was performed by applying a stage voltage to deintercalate the Li ions; the current data were collected simultaneously to evaluate the state of charge. The TEM images and EELS spectra were obtained during the charging/discharging processes. The backgrounds of the EELS spectra were removed on the basis of a polynomial function using the program DigitalMicrograph. Further, the intensities of the spectra were normalized before all analyses.

### *4.2.5 First Principle Calculations*

All the DFT calculations were performed using the Vienna ab initio simulation package (VASP) within the projector augmented wave (PAW) approach. The spin-polarized generalized gradient approximation (GGA) was employed to investigate the structures; the Perdew–Burke–Ernzerhof exchange–correlation functional was applied. The energy cutoff for the plane waves was set to 600 eV, and a Monkhorst—Pack scheme $4 \times 2 \times 2$ *k*-points mesh was used for integration in the irreducible Brillouin zone. The results were obtained within the GGA + U scheme using the following Hubbard U values: Ru (5 eV), Co (4.91 eV), Mn (4.95 eV), Ni (6.30 eV), Ti (2.5 eV), Zr (2.0 eV), and Sn (3.5 eV).

## 4.3 Results and Discussions

### *4.3.1 General Characterization of $Li_2RuO_3$ Electrode*

We chose $Li_2RuO_3$ as the model lithium-rich compound, because it contains a single metal, making it convenient to track the electronic and geometric structural changes during charging/discharging process without interference. In addition, ruthenium ions have little possibility hoping to lithium layer as observed in previous work, providing a stable environment to study. $Li_2RuO_3$ powders were prepared by a simple solid-state reaction involving the sintering of a uniform mixture of $RuO_2•H_2O$ and $Li_2CO_3$. The general characteristics of $Li_2RuO_3$ material are shown in Fig. 4.1. The XRD pattern presents relatively sharp peaks, indicative of good crystalline of the material. It can be indexed to a layered structure with C2/c symmetry where Ru layer and Li layer atoms stack alternately and extra Li ions contain in the Ru layer (Fig. 4.1a and the inset in Fig. 4.1a). The $Li_2RuO_3$ electrode was cycled under a current density of 30 mA $g^{-1}$ between 2.0 and 4.6 V, and the results are shown in Fig. 4.1b, c. Although its theoretical capacity, estimated through the redox reaction of $Ru^{4+}/Ru^{5+}$, was 164 mA h $g^{-1}$, it still delivers a specific capacity of more than 220 mA h $g^{-1}$ after 30 cycles under deep charge/discharge, showing a better stability. The voltage platform at 4.25 V for the first charge may be contributed to the oxygen

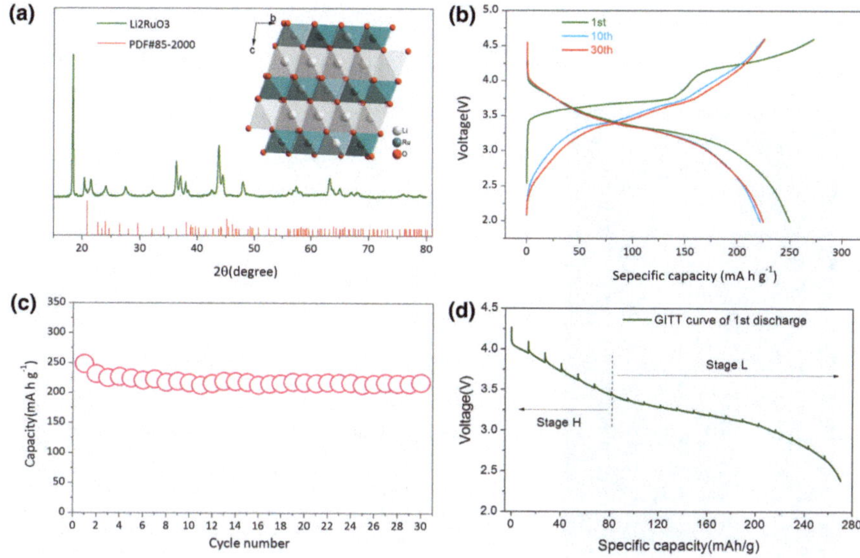


**Fig. 4.1** **a** XRD pattern of $Li_2RuO_3$ indexed using the standard spectrum of PDF#85-2000. **b** Charge/discharge curves for the first three cycles at 30 mA $g^{-1}$ between 2.0 and 4.6 V. **c** Cycling performance of $Li_2RuO_3$ at 30 mA $g^{-1}$ between 2.0 and 4.6 V. **d** GITT curve of the first discharge of $Li_2RuO_3$ electrode

redox as studied from the previous investigation [25, 26]. The oxygen redox could be responsible for the extra capacity demonstrated in the next cycles. Figure 4.1d shows the galvanostatic intermittent titration technique (GITT) curve of the first discharge process. It presents a distinctive feature for the high voltage stage (Stage H) and low voltage stage (Stage L), as labeled in the figure respectively. The GITT curve can be used to study the kinetic property of the charge/discharge processes, which is related to the certain redox process [27]. Hence, the different overpotential for Stage H and Stage L may be correlated to different redox processes, which will be characterized by other techniques.

### 4.3.2 *In Situ XRD Study*

In situ XRD technique was employed to monitor the evolution of the long-range structure order of $Li_2RuO_3$ during charging/discharging (Fig. 4.2). To demonstrate clearly the process of structure change, the XRD patterns (right side of Fig. 4.2) were translated into contour plots with the color depth representing the diffraction intensity (left side of Fig. 4.2). Before 3.8 V, it shows that there is three-phase transition during the first charge. On further oxidation to about 4.25 V plateau, which is the activation process of $Li_2RuO_3$, a solid-solution reaction with very small lattice change occurs.

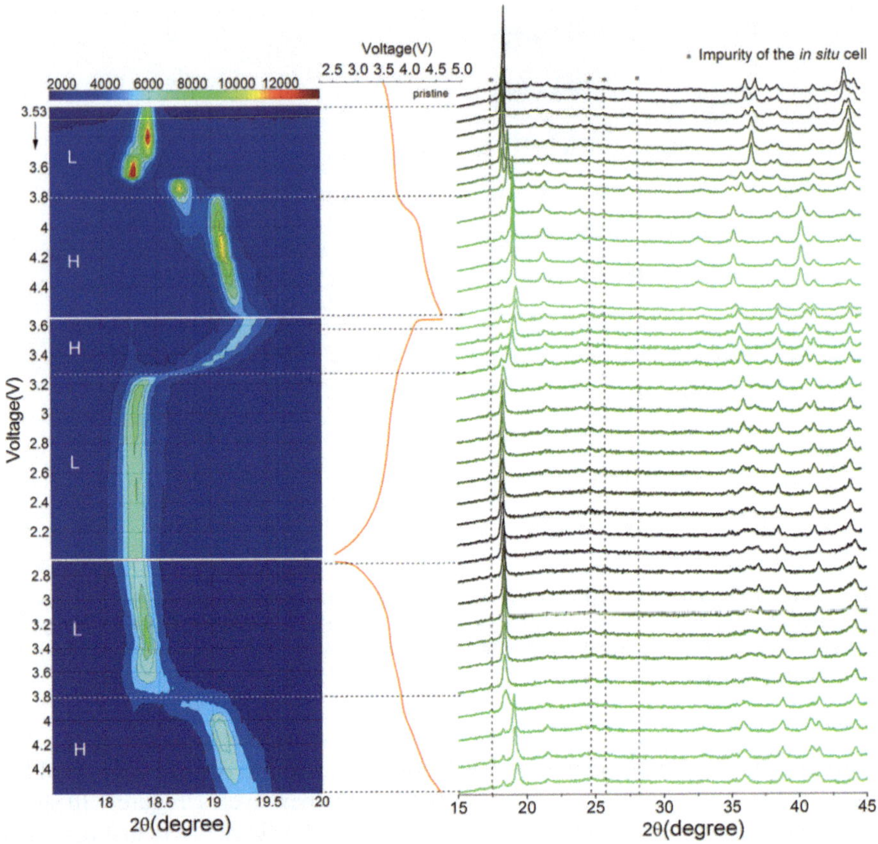


**Fig. 4.2** In situ XRD analysis of $Li_xRuO_3$ during the first charge/discharge and second charge processes

During next discharge and second charge, we can observe that two-phase change happens at certain voltage (about 3.3 for discharge and 3.8 V for charge). This phase change divided the total process into two distinctive processes, just consistent with previous part, called Stage H and Stage L. In each phase, the lattice change obeys to a solid-solution process. Unlike conventional cases, such distinct characteristics of the structural evolution of the two stages are supposed to be caused by different redox processes, which account for the high capacity referred before. Furthermore, the reversible discharge/charge profiles between the first discharge and the second charge demonstrate a good structure stability, and we did not observe any trace of spinel phase transformation.

### 4.3.3 In Situ Ru K-Edge XAFS Study

In situ Ruthenium K-edge X-ray absorption fine structure (XAFS) spectrum was measured to track the electronic structural and local coordination structural change of $Li_2RuO_3$ electrode during charge/discharge process, as shown in Fig. 4.3. Figure 4.3a

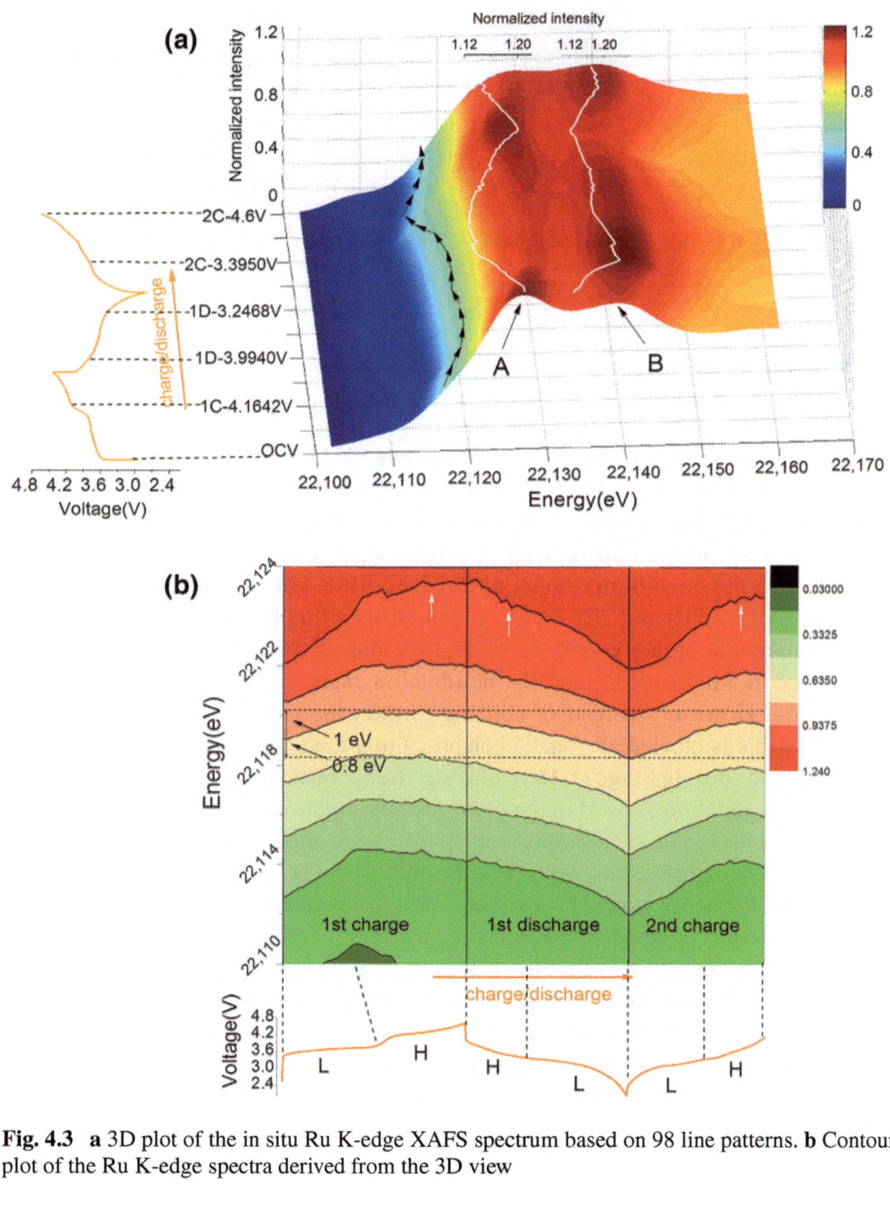


**Fig. 4.3** **a** 3D plot of the in situ Ru K-edge XAFS spectrum based on 98 line patterns. **b** Contour plot of the Ru K-edge spectra derived from the 3D view

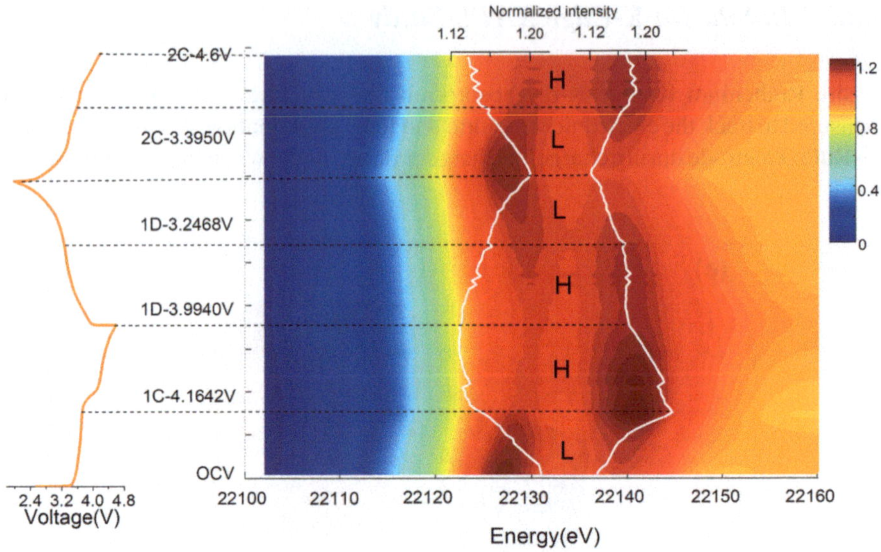


**Fig. 4.4** Bottom view of the 3D plot of Ru K-edge XAFS spectrum

demonstrates the 3D plot of the in situ Ru K-edge XAFS spectrum based on 98 line patterns. The intensity plots of peak A and B (white lines in Figs. 4.3a and 4.4) evolve reversely during the two stages, and reversible "fluctuations" are observed during Stage H but not during Stage L. Figure 4.3b further shows the contour plot derived from the three-dimensional plots of the Ru K-edge XAFS spectra between the energy of 22,110 and 22,124 eV. It can be seen that the absorption edge shifts continuously to a higher energy level by 1 eV during Stage L; this corresponds to the change of $Ru^{4+}$ to $Ru^{5+}$ and the delithiation capacity of 164 mA h $g^{-1}$ of one lithium ion. However, instead of shifting further, the absorption edge shifts back gradually (this is labeled as "back" behavior) to a lower energy level accompanied by fluctuations during Stage H. Here that the "back" behavior is a continuous change rather than a saltation viewed in the previous investigation [25, 26]. As the intensity of peak A reflects the local empty density of states, the "fluctuations" and "back" behavior of adsorption edge, which corresponds to the 1s $\rightarrow$ 5p transition, show that the density of states at stage H is related to distortion, in addition to the redox, while it is a monotonic change during stage L.

It can be seen (Fig. 4.5) that the intensity of the pre-edge peak during Stage H increases and the pre-edge is weak and does not change during Stage L. XANES structures originate from multiple scattering processes of the photoelectrons with neighboring atoms, revealing the electronic structure and the local arrangement of atoms around the central absorbing atom. Based on the dipole transition rules, the transition of the 1s electrons of Ru to the 4d states, corresponding to the pre-edge peak, is forbidden. However, when the 4d states involve the 5p states owing to the distortion of the octahedron, the dipole transition of the 1s electrons to the 5p states is

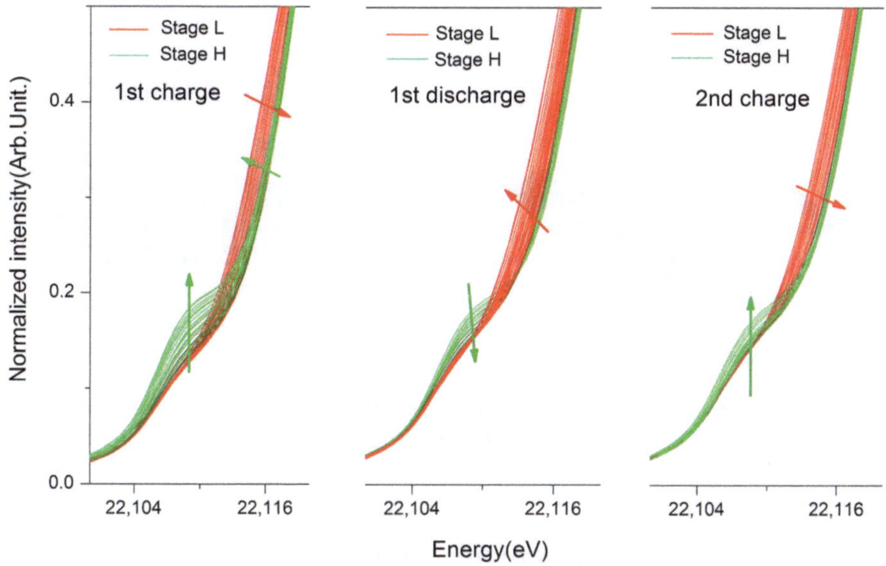


**Fig. 4.5** Evolution of the pre-edge peak of the Ru K-edge XAFS spectra during the first charge/discharge and second charge processes

allowed and results in a prominent pre-edge peak. Therefore, the increase in intensity of the pre-edge peak during Stage H could be said to be attributable to the distortion of the local structure. In contrast, during Stage L the pre-edge is weak and does not change, meaning that the redox reaction of Ru just leads to the mild shrinking of the local structure.

Moreover, the Fourier-transformed EXAFS patterns, analyzed via fitting the spectra yielding Debye–Waller factor and bond length, are shown to facilitate the analysis. As can be seen from the 3D plot of the radial distribution function (Fig. 4.6a) and its bottom view (Fig. 4.6b), the evolution of the first peak, corresponding to Ru–O coordination, exhibits contrasting trends during Stage L and Stage H. Figure 4.7 gives the fitted results of first shell of EXAFS spectra. In view of that the change of coordination number during cycling is negligible, the intensity, in direct proportion to Debye–Waller factor, reflects the disorder of the first shell environment directly. The downhill trend of the Debye–Waller factor before 3.8 V in Stage L implies that the first shell structure tends to be more orderly during delithiation. After 3.76 V, the value of Debye–Waller factor rises to high level consecutively before goes down when the voltage comes to 4.6 V, showing a gradual disorder evolution in Stage H. Furthermore, the change of fitted average bond length (Fig. 4.7b) of Ru–O presents the similar distinctive variation in the two stages, in perfect agreement with previous edge shifting analysis. All of these tell us that the local structure first shrinks orderly in Stage L and then varies to be much distorted in Stage H. This will give us the hint for the stability as discussed later.

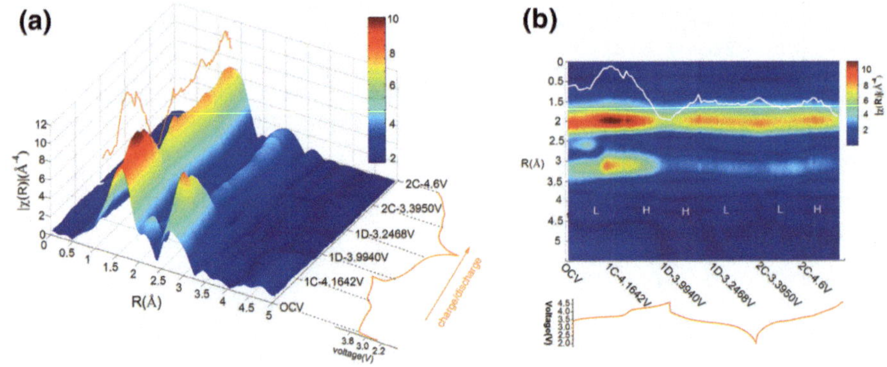


**Fig. 4.6** **a** 3D plot of the k3-weighted Fourier transform magnitudes of the in situ Ru K-edge extended X-Ray absorption fine structure (EXAFS) spectrum. **b** Bottom view of the 3D plot of radial distribution function patterns

### *4.3.4 In Situ EELS Study of the Electronic Structure of O*

To determine the intrinsic mechanism of Stage H, in situ transmission electron microscopy (TEM) and electron energy loss spectroscopy (EELS), which do not involve the side reaction owing to the binder and the liquid organic electrolyte, were employed to further study the changes in the electronic structure of the O ions directly. A nanobattery was fabricated using $Li_2RuO_3$ particles adhering to the tip of a tungsten pin as the cathode, a gold tip dipped with lithium metal as the anode, and the $Li_2O$ on the surface of lithium as the solid electrolyte, as the schematic (Fig. 4.8a) illustrates. The sample of $Li_2RuO_3$ electrode is newly prepared, and the surface is clean. Constant bias voltage is given on the cathode side, with the other grounding connection. Charging and discharging processes for the first cycle are realized by controlling the voltage to a proper value for a certain time, during which the variation of the current was recorded to estimate the state of charge (Fig. 4.9). The in situ TEM images show that there was no obvious change in the shape of nanobattery during charge/discharge process. The EELS attached on TEM accessed by more than 100 k eV electrons can provide bulk information rather than the surface information as demonstrated in previous work [19]. In addition, high vacuum and the absence of electrolyte can keep the collected information from the interference of other species like $Li_xO_2$, $Li_2CO_3$. O K-edge EELS spectra were collected to monitor the electronic structural change upon charging and discharging (Fig. 4.8b), with specific positions on the $Li_2RuO_3$ particle where the EELS spectra were taken, as shown by the color dots in Fig. 4.8c. Spectra A and B were collected during Stage L, while C and D were taken during Stage H, as predicted in Table 4.1. In view of that the pre-edge peak of O K-edge spectra corresponds to transition of O 1s electrons to hybridized states of O 2p and Ru 4d orbitals, the integrated area shown in Fig. 4.8d can be deemed into the indicator of the oxidation state of O ion. It can be seen that there is little change

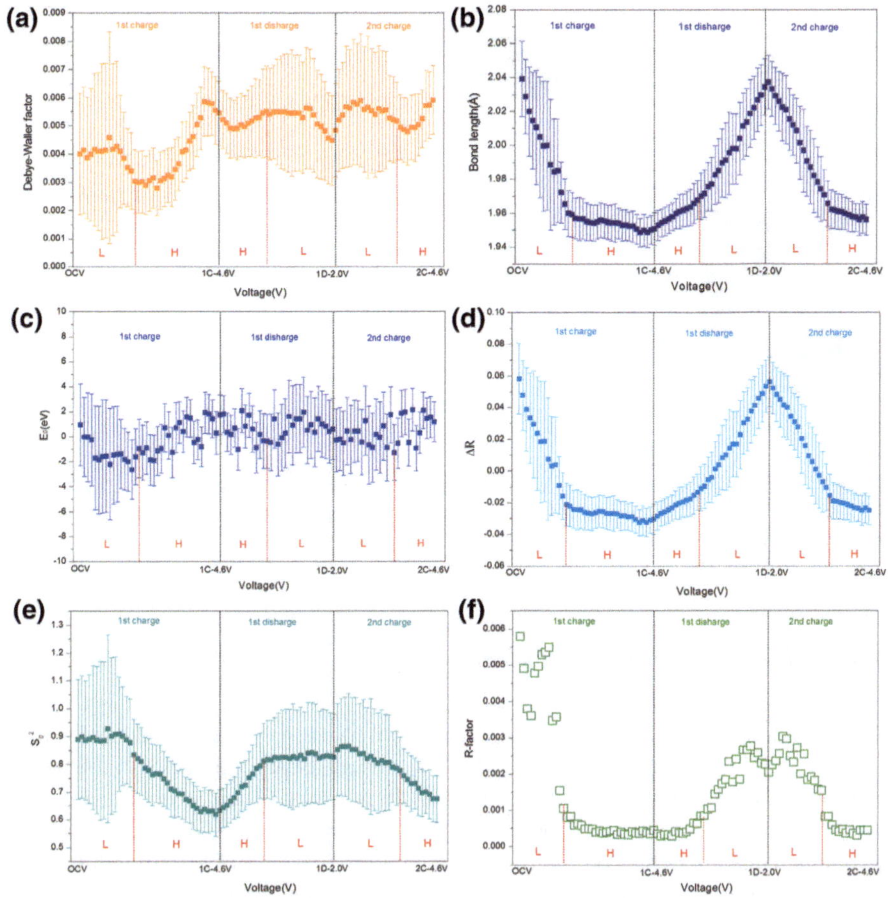


**Fig. 4.7** Evolution of the EXAFS fitting parameters. **a** the Debye–Waller factor, **b** fitted bond length, **c** the energy origin displacement parameter, **d** bond length parameter used for the fitting, **f** the R-factors for the fitting

of the pre-edge peak in Stage L, while the pre-edge peak of the O K-edge EELS spectra changes significantly and shifts to a higher energy level for approximately 0.5 eV during Stage H, in keeping with previous reports on the appearance of $O_2^{2-}$ species [25]. When discharging the material to the E state, the edge returned to the original position and the peak area shrank, indicating that the redox process during Stage H is attributable mainly to the O ions and reversible during charge/discharge process.

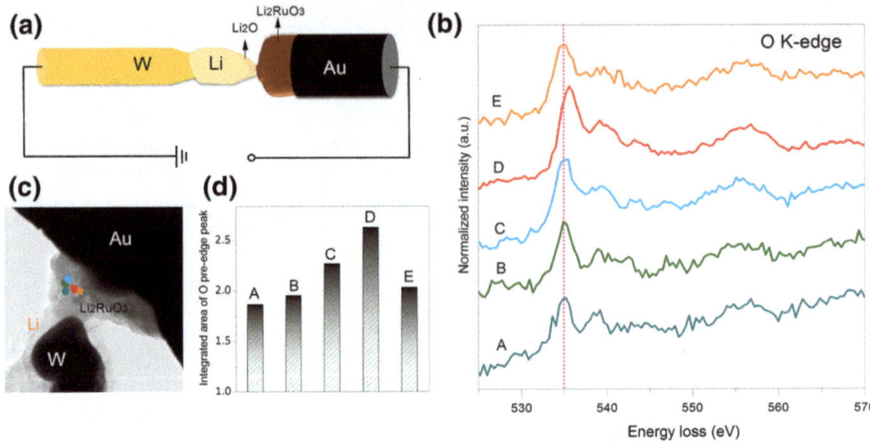


**Fig. 4.8** **a** Schematic of the nanodevice for in situ TEM and EELS measurements. **b** EELS spectra of O K-edge. **c** Real TEM image of the nanobattery with positions of the EELS spectra collected are marked by dots colored correspondingly to the spectra in **b** and **c**. **d** Integrated area calculated from the pre-edge peak of O K-edge EELS spectra

### *4.3.5 First Principle Calculations*

Similar anionic redox participation has been observed in conventional oxides (e.g., $LiCoO_2$, $LiNiO_2$, and $LiNi_{1/3}Co_{1/3}Mn_{1/3}O_2$) using experimental and theoretical methods [28–34]. However, these materials do not retain their structural integrity, owing to the holes generated on the 2p band of the O ions. Therefore, the high reversible capacity exhibited by lithium-rich compound $Li_2RuO_3$ with similar layer structure seems to be counterintuitive. Density functional theory (DFT) calculations were performed to confirm the fine structural changes observed. Lithium removal of $Li_{16}Ru_8O_{24}$ supercell is performed one by one till $Ru_8O_{24}$ is reached after determining the delithiation sequence of the two different Li sites (Fig. 4.10). During the delithiation, the total structure remains to be stable and no metal atoms hooping happen. As shown in Fig. 4.11a, during Stage L ($x \geq 8$), the local structure of $Li_xRu_8O_{24}$ in the transition metal layer remains intact along the c-axis view. In contrast, during Stage H ($x \leq 8$), the $RuO_6$ octahedrons show large distortions. This is consistent with the results of the previous pre-edge analysis performed on the basis of the Ru K-edge XAFS spectra. Moreover, the length of the O–O bond of the edge shared by two octahedrons reduces gradually, from 2.710 to 2.418 Å, resulting in the formation of $O_2^{2-}$ species. All the Ru–O bonds decrease converging to a lower length of about 1.98 Å in Stage L, but diverge to quite dispersive values in Stage H (Fig. 4.11b). This is consistent with previous findings that Stage L is dominated by redox $Ru^{4+}/Ru^{5+}$ inducing structural change elegantly, while in Stage H anionic redox is activated and causes great distortion. All of Ru–O bonds can be divided into two types: short bonds (bonds A, green color in Fig. 4.11b) and long bonds (bonds B, red color in Fig. 4.11b).

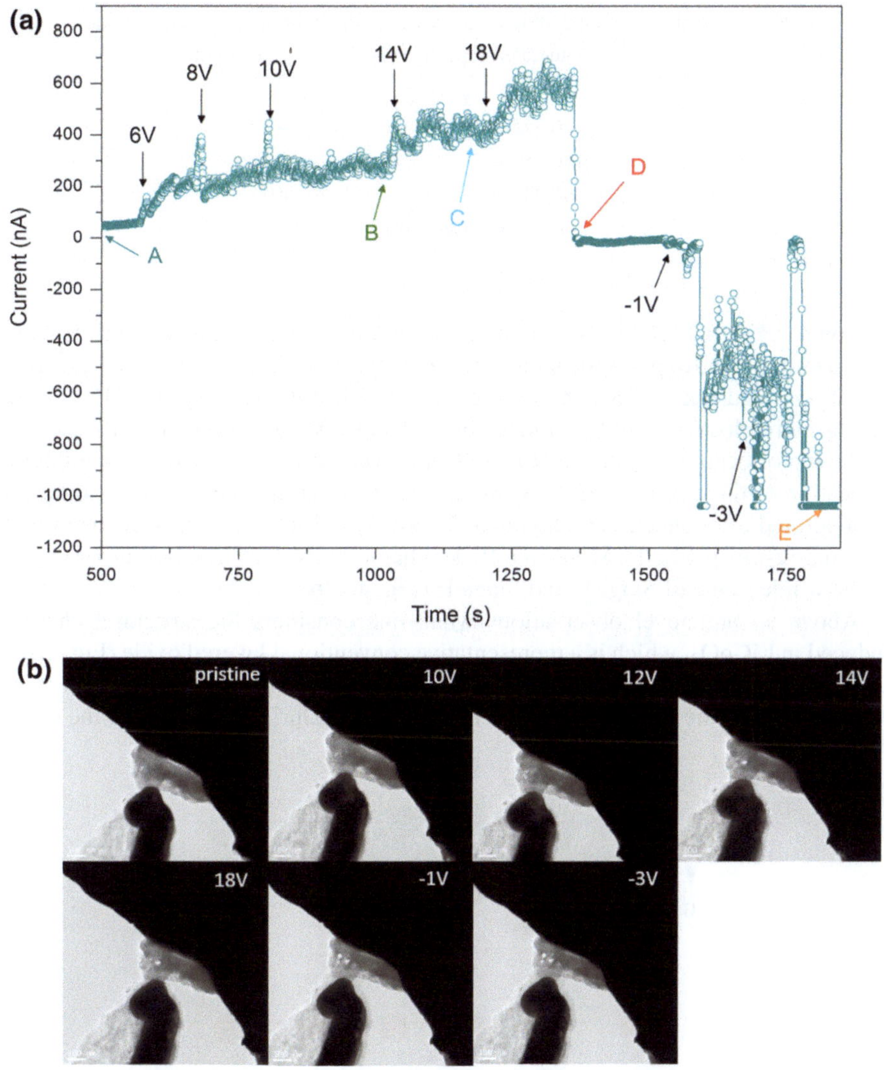


**Fig. 4.9** **a** Current data for in situ charge and discharge by adding staged voltage as shown in the figure. **b** In situ TEM images of the nanobattery device

It is the competition between the two types of bonds that induce the fluctuations of the electronic structural of Ru in Stage H: The A bonds tend to shrink, causing the Ru ions to become more oxidized, while the B bonds tend to stretch, causing the Ru ions to become more reduced, demonstrating the fluctuation of oxidation state of Ru during Stage H.

The electronic structural change is also reflected by density of states (DOS) analysis directly. From the PDOS results of Ru and O (Fig. 4.11c, d, total DOS analysis

**Table 4.1** Stage of charge corresponding to the times at which the spectra were collected

| Spectra | Integrated capacity (nA·s) | State of charge |
|---|---|---|
| A | 0 | 0 |
| B | 126,975 | 47.5% |
| C | 186,688 | 70.0% |
| D | 266,942 | 100.0% |
| E | −156,956 (discharge) | 58.6% |

is shown in Fig. 4.12), we find that the electrons are donated from both of Ru and O, but quantitatively Ru plays the major role, as suggested from the integrated electrons of the occupied states of Ru and O (Fig. 4.13a). When it comes to Stage H, O starts giving more electrons while Ru takes little change. Most importantly, the energies of the top of occupied states of Ru and O are lowered when $x = 4$ and 0, implying relocation of the electrons of the system. The "fluctuating" and "back" feature can be identified after check the integrated DOS of Ru (Fig. 4.13a), in agreement with previous results perfectly. Moreover, Bader charge analysis shows the similar feature in the whole range of Stage L and Stage H (Fig. 4.13b).

Above all are novel observations when one considers the structural changes induced in $LiCoO_2$, which is a representative conventional layered oxide (Fig. 4.14). Owing to the anionic redox reaction of O, the $CoO_6$ octahedron maintains its structure throughout the process, in contrast to $Li_2RuO_3$. During delithiation, the Co–O

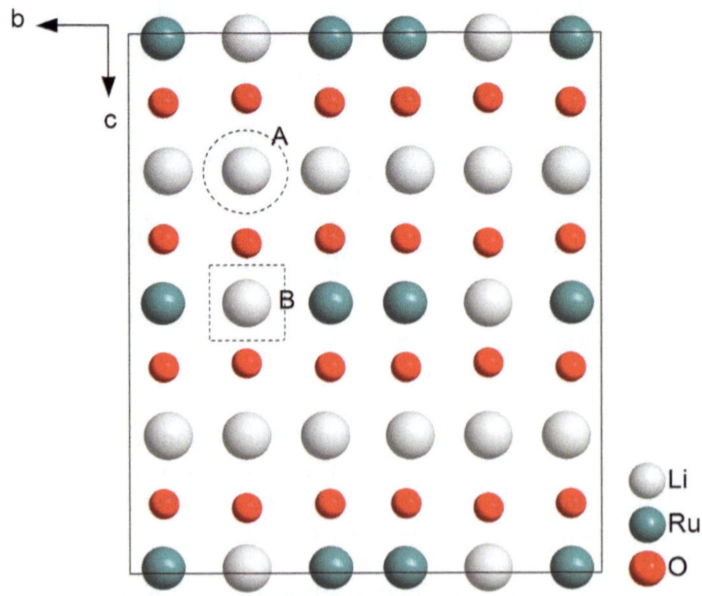


**Fig. 4.10** Determination of the delithiation sequence of the two Li sites

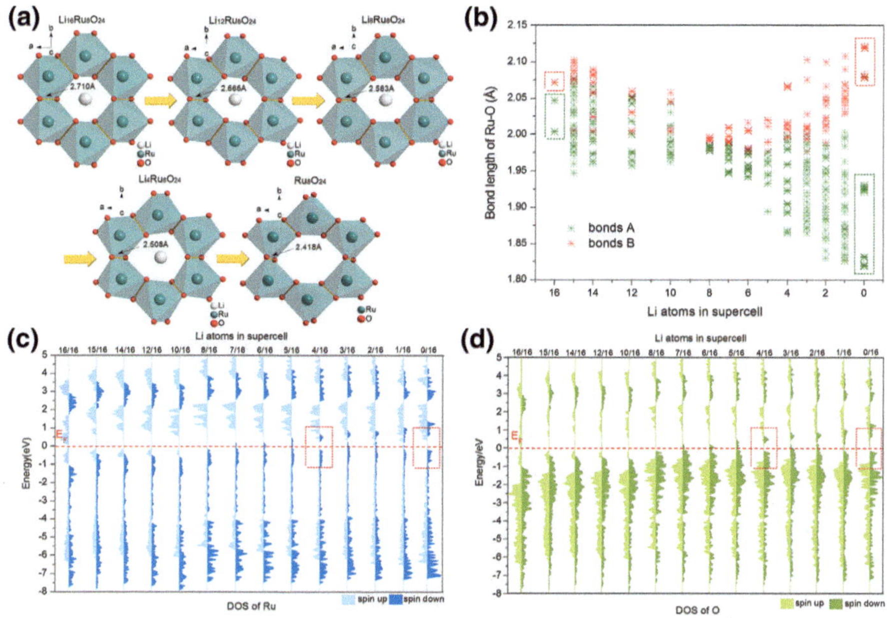


**Fig. 4.11** Results of DFT calculations of $Li_xRuO_3$ during delithiation. **a** Evolution of the local structure of $Li_xRuO_3$ from the viewpoint of the transition metal layer, **b** evolution of the Ru–O bond length in $Li_xRuO_3$. **c**, **d** evolution of the density of state (DOS) structure of Ru and O in $Li_xRuO_3$

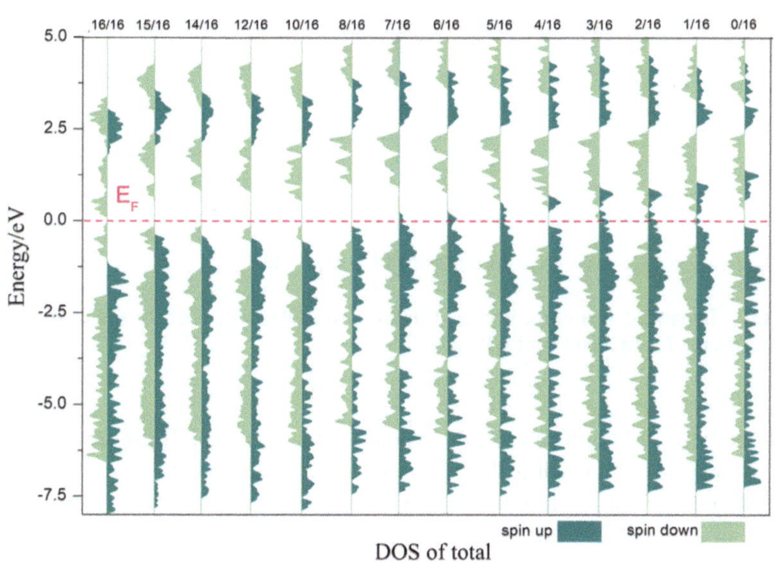


**Fig. 4.12** Evolution of the total DOS of $Li_xRu_8O_{24}$

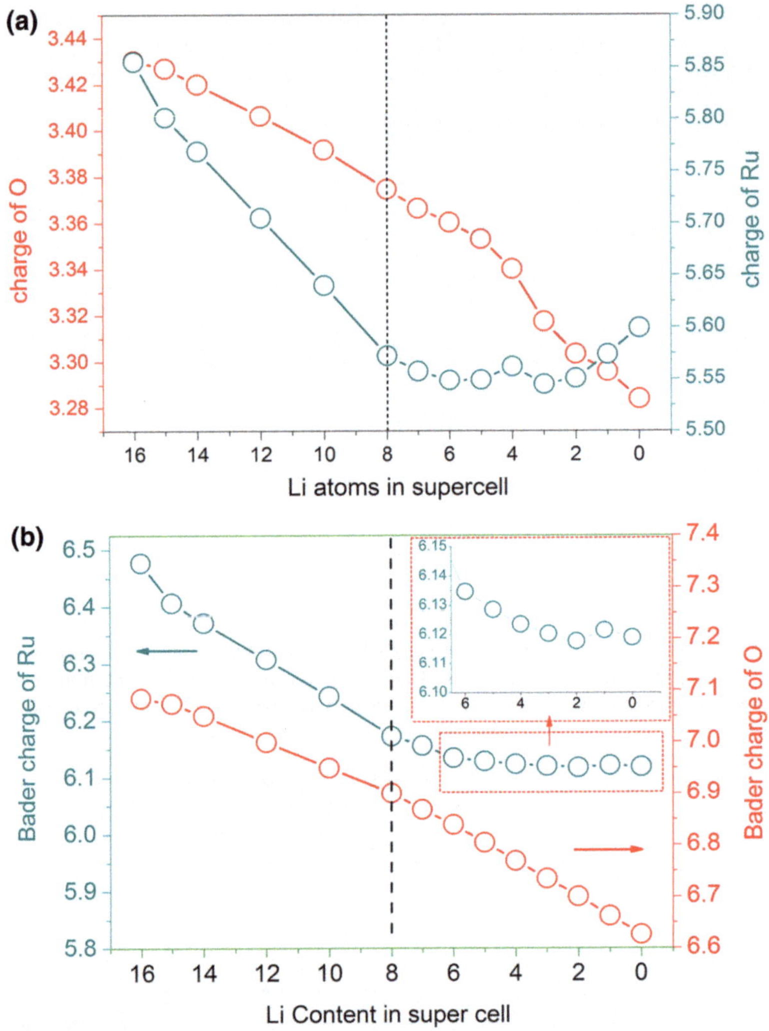


**Fig. 4.13** **a** Integrated area of the occupied states for the DOS values of Ru and O. **b** Analysis of the Bader charge for Ru and O in $Li_xRu_8O_{24}$

bonds decrease in length to 1.87 Å, instead of increasing. All the bonds are in the range of 1.867–1.996 Å, while the bonds of $Li_2RuO_3$ are in the range of 1.821–2.121 Å, which means that the amplitude of bonds variation in $LiCoO_2$ is much smaller than that in $Li_2RuO_3$. The difference between $LiCoO_2$ and $Li_2RuO_3$ is so inexplicable since their structure is analogous except for one thing the superstructure in the transition metal layer. As shown in Fig. 4.15, during Stage H, the Li-driven process in $Li_2RuO_3$ is mainly charge-compensated by oxygen ions, resulting in the pulling of the two oxygen atoms closer; this phenomenon is aided by the fact that the Li atoms

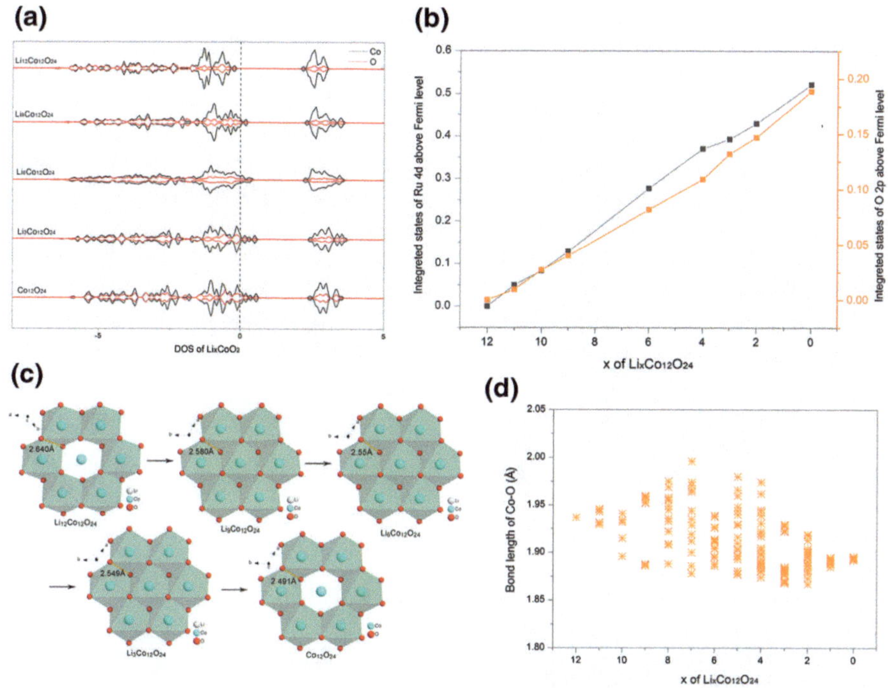


**Fig. 4.14** Analysis of electronic structure of Li*x*CoO$_2$ during delithiation. **a** The evolution of the DOS of the Co and O of Li*x*CoO$_2$ when $x = 12, 9, 6, 3$, and 0. The integrated areas of the states that cross the Fermi level of Co and O are shown in **b** to quantitatively describe the charge compensation mechanism. **c** Local structural change of the transition metal layer of LiCoO$_2$. **d** Evolution of the Co–O bond during the delithiation of LiCoO$_2$

in the transition metal layer have almost no Coulombic effect on the O atoms. In the end, a pseudo-$O_2^{2-}$ species is formed, and the $RuO_6$ octahedron undergoes distortion to lower the energy of the structure. This causes fluctuations in the valence of the Ru ions. However, in the case of $LiCoO_2$, the O–O bond does not shrink to form $O_2^{2-}$ species, owing to the large Coulombic effect of Co on the coordinated O atoms (Fig. 4.15d). Thus, the resulting rigid, highly homogeneous structure of $LiCoO_2$ does not allow for structural stabilization, in contrast to $Li_2RuO_3$. As a result, the Co–O bond breaks when stretched by the large force attributable to the changes in the electronic structure. This leads to a phase transition, decomposition, and oxygen release of $LiCoO_2$ [35]. Therefore, the $LiCoO_2$ can only deliver practical specific capacity of about 145 mA h g$^{-1}$. This means the superstructure in $Li_2RuO_3$ provides a flexible feature, leading to higher oxygen stability. Figure 4.16 further shows the calculated oxygen vacancy formation enthalpy ΔH as a function of lithium content in the supercell of $Li_{16}Ru_8O_{24}$ and $Li_{12}Co_{12}O_{24}$. The ΔH for $Li_2RuO_3$ and $LiCoO_2$ was defined as $\Delta H(x) = E(Li_xRu_8O_{23}) + 1/2E(O_2) - E(Li_xRu_8O_{24})$ and $\Delta H(x) = E(Li_xCo_{12}O_{23}) + 1/2E(O_2) - E(Li_xCo_{12}O_{24})$, respectively. The ΔH of $LiCoO_2$ is

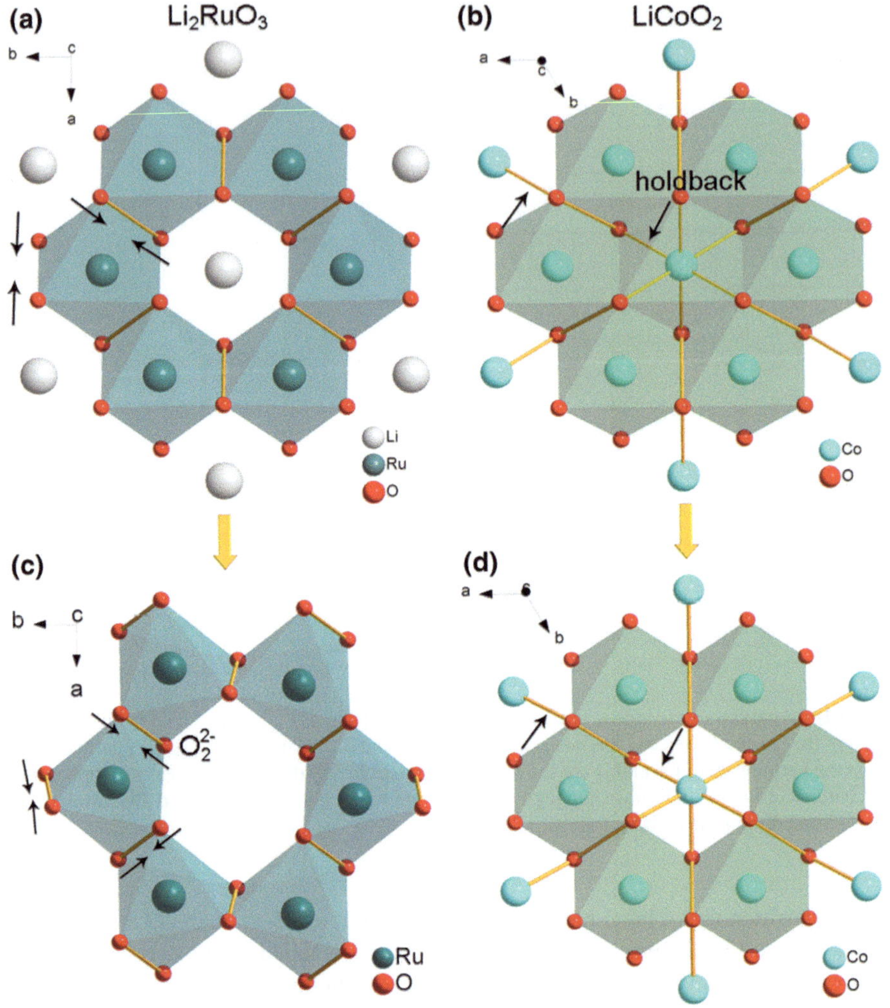


**Fig. 4.15** **a** Honeycomb structure view of the transition metal layer of $Li_2RuO_3$. **b** Local structural view of the transition metal layer of $LiCoO_2$. **c** Honeycomb structure view of the transition metal layer of totally delithiated $Li_2RuO_3$. **d** Local structural view of the transition metal layer of totally delithiated $LiCoO_2$

below 0 after half of the lithium ions are deintercalated, suggesting that oxygen atoms are not stable after deeply delithiated. However, in the case of $Li_2RuO_3$, the $\Delta H$ of $Li_xRu_8O_{24}$ is over 0 even when totally delithiated, demonstrating that the formation of oxygen vacancy is thermodynamically unfavorable. This flexible structure providing by the superlattice facilitating the oxygen redox is also confirmed by performed DFT calculations relating to the structure of $Li_2MO_3$ (M = Ti, Sn, Mo, and Zr) and $LiMO_2$ (M = Ni and $Ni_{1/3}Co_{1/3}Mn_{1/3}$) (Fig. 4.17).

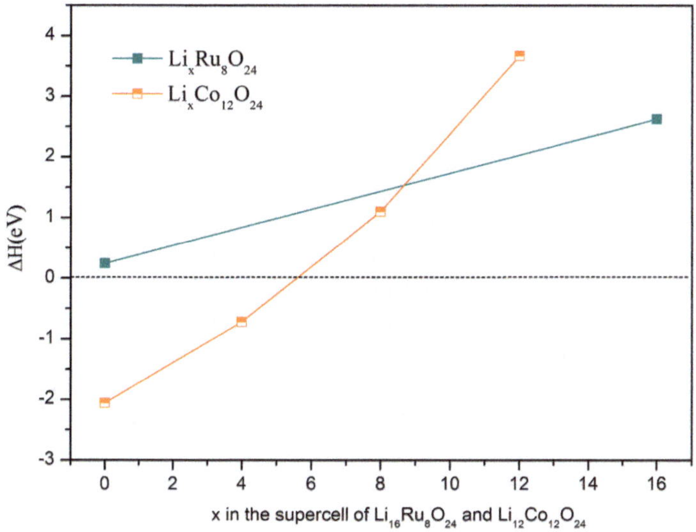


**Fig. 4.16** Calculated oxygen vacancy formation enthalpy ΔH as a function of lithium content in the supercell of $Li_{16}Ru_8O_{24}$ and $Li_{12}Co_{12}O_{24}$

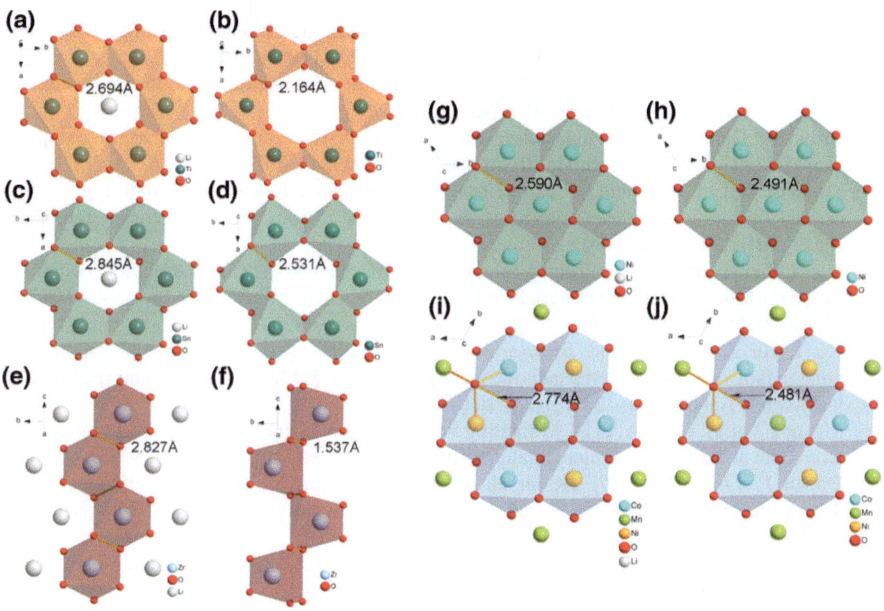


**Fig. 4.17** Local structural change before and after delithiation of several $Li_2MO_3$ and $LiMO_2$ oxides. **a** $Li_2TiO_3$, **b** delithiated $Li_2TiO_3$, **c** $Li_2SnO_3$, **d** delithiated $Li_2SnO_3$, **e** $Li_2ZrO_3$, **f** delithiated $Li_2ZrO_3$, **g** $LiNiO_2$, **h** delithiated $LiNiO_2$, **i** $LiNi_{1/3}Co_{1/3}Mn_{1/3}O_2$, **j** delithiated $LiNi_{1/3}Co_{1/3}Mn_{1/3}O_2$

## 4.4 Summary

In this work, we explain that why lithium-rich materials $Li_2RuO_3$ can supply such high capacities rather than conventional layered oxides $LiCoO_2$. During charge/discharge process, $Li_2RuO_3$ undergoes two different redox processes. At Stage L, the $Li_2RuO_3$ electrode can deliver 164 mA h $g^{-1}$ through the redox reaction of $Ru^{4+}/Ru^{5+}$. At Stage H, a pseudo-$O_2^{2-}$ species is formed, and the $RuO_6$ octahedron undergoes distortion to lower the energy of the structure, facilitated by the flexible structure provided by lithium atoms in the transition metal layer. This induces a higher oxygen stability, favors the structural integrity, and provides high-specific capacity, which can be applied in explaining the difference of structural stability between conventional layered oxides and lithium-rich cathodes. This work would facilitate the developing of higher capacity electrodes for lithium-ion batteries.

## References


1. Padhi AK, Nanjundaswamy KS, Masquelier C, et al. Effect of structure on the Fe3+/Fe2+ redox couple in iron phosphates. J Electrochem Soc. 1997;144(5):1609–13.
2. Ohzuku T, Ueda A, Nagayama M. Electrochemistry and structural chemistry of $LiNiO_2$ ($R\bar{3}m$) for 4 v secondary lithium cells. J Electrochem Soc. 1993;140(7):1862–70.
3. Gu M, Belharouak I, Zheng J, et al. Formation of the spinel phase in the layered composite cathode used in Li-ion batteries. ACS Nano. 2013;7(1):760–7.
4. Croy JR, Gallagher KG, Balasubramanian M, et al. Examining hysteresis in composite $xLi_2MnO_3·(1-x)LiMO_2$ cathode structures. J Phys Chem C. 2013;117(13):6525–36.
5. Mohanty D, Li J, Abraham DP, et al. Unraveling the voltage-fade mechanism in high-energy-density lithium-ion batteries: origin of the tetrahedral cations for spinel conversion. Chem Mater. 2014;26(21):6272–80.
6. Lyu Y, Zhao N, Hu E, et al. Probing reversible multielectron transfer and structure evolution of $Li_{1.2}Cr_{0.4}Mn_{0.4}O_2$ cathode material for Li-ion batteries in a voltage range of 1.0–4.8 V. Chem Mater. 2015;27(15):5238–52.
7. Xu Y, Hu E, Yang F, et al. Structural integrity—searching the key factor to suppress the voltage fade of Li-rich layered cathode materials through 3D X-ray imaging and spectroscopy techniques. Nano Energy. 2016;1(28):164–71.
8. Zheng J, Gu M, Genc A, et al. Mitigating voltage fade in cathode materials by improving the atomic level uniformity of elemental distribution. Nano Lett. 2014;14(5):2628–35.
9. Dogan F, Long BR, Croy JR, et al. Re-entrant lithium local environments and defect driven electrochemistry of Li- and Mn-rich Li-ion battery cathodes. J Am Chem Soc. 2015;137(6):2328–35.
10. Koga H, Croguennec L, Ménétrier M, et al. Operando X-ray absorption study of the redox processes involved upon cycling of the Li-rich layered oxide $Li_{1.20}Mn_{0.54}Co_{0.13}Ni_{0.13}O_2$ in Li ion batteries. J Phys Chem C. 2014;118(11):5700–9.
11. Koga H, Croguennec L, Ménétrier M, et al. Different oxygen redox participation for bulk and surface: A possible global explanation for the cycling mechanism of $Li_{1.20}Mn_{0.54}Co_{0.13}Ni_{0.13}O_2$. J Power Sources. 2013;236:250–8.
12. Luo K, Roberts MR, Hao R, et al. Charge-compensation in 3d-transition-metal-oxide intercalation cathodes through the generation of localized electron holes on oxygen. Nat Chem. 2016;8(7):684–91.

13. Muhammad S, Kim H, Kim Y, et al. Evidence of reversible oxygen participation in anomalously high capacity Li- and Mn-rich cathodes for Li-ion batteries. Nano Energy. 2016;21:172–84.
14. Sun YK, Lee MJ, Yoon CS, et al. The role of $AlF_3$ coatings in improving electrochemical cycling of Li-enriched nickel-manganese oxide electrodes for Li-ion batteries. Adv Mater. 2012;24(9):1192–6.
15. Wu F, Li N, Su Y, et al. Spinel/layered heterostructured cathode material for high-capacity and high-rate Li-ion batteries. Adv Mater. 2013;25(27):3722–6.
16. Qiao Q Q, Zhang H Z, Li G R, et al. Surface modification of Li-rich layered $Li(Li_{0.17}Ni_{0.25}Mn_{0.58})O_2$ oxide with $LiMnPO_4$ as the cathode for lithium-ion batteries. J Mate Chem A. 2013;1(17):5262.
17. Liu W, Oh P, Liu X, et al. Countering voltage decay and capacity fading of lithium-rich cathode material at 60 & #xB0;C by hybrid surface protection layers. Adv Energy Mater. 2015;5(13):1500274.
18. Zhang X, Belharouak I, Li L, et al. Structural and electrochemical study of $Al_2O_3$ and $TiO_2$ coated $Li_{1.2}Ni_{0.13}Mn_{0.54}Co_{0.13}O_2$ cathode material using ALD. Adv Energy Mater. 2013;3(10):1299–1307.
19. Pei Y, Xu C-Y, Xiao Y-C, et al. Phase transition induced synthesis of layered/spinel heterostructure with enhanced electrochemical properties. Adv Functional Mater. 2017;1604349.
20. Zhao Y, Liu J, Wang S, et al. Surface structural transition induced by gradient polyanion-doping in Li-rich layered oxides: Implications for enhanced electrochemical performance. Adv Functional Mate. 2016;26(26):4760–7.
21. Zheng F, Yang C, Xiong X, et al. Nanoscale surface modification of lithium-rich layered-oxide composite cathodes for suppressing voltage fade. Angew Chem Int Ed. 2015;127(44):1–6.
22. Bian X, Fu Q, Qiu H, et al. High-performance $Li(Li_{0.18}Ni_{0.15}Co_{0.15}Mn_{0.52})O_2@Li_4Mn_5O_{12}$ heterostructured cathode material coated with a lithium borate oxide glass layer. Chem Mater. 2015;27(16):5745–54.
23. He W, Yuan D, Qian J, et al. Enhanced high-rate capability and cycling stability of Na-stabilized layered $Li_{1.2}[Co_{0.13}Ni_{0.13}Mn_{0.54}]O_2$ cathode material. J Mater Chem A. 2013;1(37):11397.
24. Zhang HZ, Qiao QQ, Li GR, et al. $PO_4^{3-}$ polyanion-doping for stabilizing Li-rich layered oxides as cathode materials for advanced lithium-ion batteries. J Mater Chem A. 2014;2(20):7454.
25. Sathiya M, Rousse G, Ramesha K, et al. Reversible anionic redox chemistry in high-capacity layered-oxide electrodes. Nat Mater. 2013;12(9):827–35.
26. Sathiya M, Abakumov AM, Foix D, et al. Origin of voltage decay in high-capacity layered oxide electrodes. Nat Mater. 2015;14(2):230–8.
27. Jung S-K, Gwon H, Hong J, et al. Understanding the degradation mechanisms of $LiNi_{0.5}Co_{0.2}Mn_{0.3}O_2$ cathode material in lithium ion batteries. Adv Energy Mater. 2014;4(1):1300787.
28. Graetz J, Hightower A, Ahn CC, et al. Electronic structure of chemically-delithiated $LiCoO_2$ studied by electron energy-loss spectrometry. Journal Phys Chem B. 2002;106(6):1286–9.
29. Wolverton C, Zunger A. First-principles prediction of vacancy order-disorder and intercalation battery voltages in $Li_xCoO_2$. Phys Rev Lett. 1998;81(3):606–9.
30. Aydinol MK, Kohan AF, Ceder G, et al. Ab initio study of lithium intercalation in metal oxides and metal dichalcogenides. Phys Rev B. 1997;56(3):1354–65.
31. Yoon WS, Kim KB, Kim MG, et al. Oxygen contribution on Li-ion intercalation−deintercalation in $LiCoO_2$ investigated by O K-edge and Co L-edge X-ray absorption spectroscopy. J Phys Chem B. 2002;106(10):2526–32.
32. Uchimoto Y, Sawada H, Yao T. Changes in electronic structure by Li ion deintercalation in $LiNiO_2$ from nickel L-edge and O K-edge XANES. J Power Sources. 2001;97–98:326–7.
33. Yoon W-S, Balasubramanian M, Chung KY, et al. Investigation of the charge compensation mechanism on the electrochemically Li-ion deintercalated $Li_{1-x}Ni_{1/3}Mn_{1/3}Co_{1/3}O_2$ electrode system by combination of soft and hard X-ray absorption spectroscopy. J Am Chem Soc. 2005;127(49):17479–87.

34. Oishi M, Fujimoto T, Takanashi Y, et al. Charge compensation mechanisms in $Li_{1.16}Ni_{0.15}Co_{0.19}Mn_{0.50}O_2$ positive electrode material for Li-ion batteries analyzed by a combination of hard and soft X-ray absorption near edge structure. J Power Sources. 2013;222:45–51.
35. Van der Ven A, Aydinol MK, Ceder G, et al. First-principles investigation of phase stability in $Li_xCoO_2$. Phys Rev B. 1998;58(6):2975–87.

# Chapter 5
# Tuning the Reversibility of Oxygen Redox in Lithium-Rich Layered Oxides

**Biao Li, Huijun Yan, Yuxuan Zuo and Dingguo Xia**

## 5.1 Introduction

The fast development of electrical vehicle market proposes high-energy storage requirements for lithium-ion batteries. In view of that, high-capacity cathodes are predominantly crucial for the realization of high-energy-density batteries. Rock-salt layered oxides $LiMO_2$ (M = Co, Ni, $Co_xNi_yMn_z$, etc.) is a common kind of cathodes with capacities of 140–200 mA h/g, [1–3] but still cannot manage with the challenge. Lithium-rich layered manganese-based oxides $Li_{1+x}Ni_yCo_zMn_{1\text{-}x\text{-}y\text{-}z}O_2$ are the well-known high-capacity cathodes, but with deficiencies like voltage decay and capacity fading [4–7]. Though tremendous efforts were made to modify the $Li_{1+x}Ni_yCo_zMn_{1\text{-}x\text{-}y\text{-}z}O_2$ electrodes, the practical application still has a far way. Recent years, many groups have taken studies on new lithium-rich layered oxides in view of their high capacities, such as the Ru-based, Nb-based, and Mo-based lithium-rich oxides [8–17]. Tarascon's group found that the high capacity of lithium-rich layered oxides is realized by concurrent participation of cationic and anionic redox processes, based on several studies of lithium ruthenium oxides [10–12]. These lithium-rich materials all have high capacities exceeding 200 and even 300 mA h/g, exhibiting high potential for the application in high-energy-density lithium-ion batteries.

This chapter was adapted from B. Li et al. *Chemistry of Materials* 2017, 29 (7), 2811–2818 with permission. Copyright 2017, American Chemical Society.

B. Li · H. Yan · Y. Zuo · D. Xia (✉)
Beijing Key Laboratory of Theory and Technology for Advanced Batteries Materials, College of Engineering, Peking University, Beijing 100871, P. R. China
e-mail: dgxia@pku.edu.cn

B. Li
e-mail: libiao2008@pku.edu.cn

 
B. Li, *Studies on Anionic Redox in Li-Rich Cathode Materials of Li-Ion Batteries*, Springer Theses, https://doi.org/10.1007/978-981-13-2847-3_5

By now, the family of lithium-rich layered materials is abundant with various new materials proposed and opens a new research hotspot for cathode materials. However, among these lithium-rich layered oxides, tremendous efforts should be dedicated to stabilize their structure during deep charge and discharge. Due to the involving of the anionic redox, deep charge and discharge for high capacity will damage the crystal framework since anions like oxygen ions will be excessively oxidized. As a result, all of the lithium-rich materials share a common feature that the cycle stability is very poor, which highly abates their high-capacity advantage than conventional cathode electrodes. Therefore, it is crucial to enhance the stability or reversibility of the anionic redox in the lithium-rich material system.

Several works have been conducted in probing and discussing the intrinsic mechanism involving the reversibility of the oxygen redox by Tarascon's and Doublet's groups [18, 19]. They proposed a metal-driven reductive coupling mechanism to explain the partial reversibility of the oxygen redox in $Li_2RuO_3$-based oxides in finding that the metal will be reduced finally at the end of the charge process. From their perspectives, the M(3d)-O(2p) (M denotes transition metal ions) covalent or the overlap between M(3d) and O(2p) states is responsible for the triggering of the reductive coupling mechanism. However, it is hard to explain why Co is not reduced according to this mechanism when large amounts of holes are created on the 2p states of oxygen ions in $LiCoO_2$ under total delithiation, though the covalence of Co–O is strong enough [20–22]. But in our previous work, we have presented from the view of geometric structure that excess lithium ions in the transition metal layer in lithium-rich oxides will enhance the flexibility and robustness of the local structure, which will facilitate the reversibility of the oxygen redox. Therefore, the triggering of the reductive coupling mechanism may also concern with the excess lithium feature of the lithium-rich layered oxides, which well explains its absence in $LiCoO_2$ which has no additional lithium ions in the transition metal layer.

However, though the reversibility of oxygen redox can be realized by reductive coupling mechanism, the excessive oxidation of oxygen still leads to large irreversible capacity. To obtain reversible anionic redox for lithium-rich high-capacity electrodes, tuning the electronic structure between metal and ligands is an effective way. Here, on the basis of a flexible structure of lithium-rich ruthenium oxide, we present that the reversibility of the oxygen redox can be controlled by tuning the electronic structure between Ru and O, which was realized by boron doping in the interstitial sites of the layered crystal structure. The doped samples show more stable structure and present different charge and discharge behaviors as characterized by in situ XRD patterns. The results of Ru K-edge XAFS spectra demonstrate that the reductive coupling mechanism did not happen without the valence of Ru shifting back. Combined with the DFT calculation, we believe that the introduction of boron atoms changes the electronic structure of Ru–O system and lowers the energy of oxygen 2p band. The building of an M–L–X (e.g., Ru–O–B) system lowering the energy of the local structure is responsible to suppress the reductive coupling process. This work provides a general strategy to tune the reversibility of anionic redox in obtaining high-capacity electrodes and opens a new scope in designing electrodes.

## 5.2 Methods

### 5.2.1 Preparation of $Li_2RuO_3$ and $Li_2RuB_xO_3$

$Li_2RuO_3$ and B$x$-LRO ($x = 0.05$, 0.1, and 0.15. For convenience, the samples doped with different amounts of boron are designated as B0.05-LRO, B0.10-LRO, and B0.15-LRO, corresponding to theoretical atomic ratio of B/Ru of 0.05, 0.10, and 0.15, respectively) samples were prepared through a high-temperature solid-state method. Stoichiometric amounts of $RuO_2 \cdot xH_2O$ (J&K, 54% Ru), $H_3BO_3$ (J&K, 99.99%), and $Li_2CO_3$ (10, 30, 50, and 70% excess for $Li_2RuO_3$, B0.05-LRO, B0.10-LRO, and B0.15-LRO, respectively. Sinopharm, 99.99%) were mixed together and ground homogeneously after being removed the crystal water under 300 °C for 6 h. The mixture was then transferred to a silica tube furnace and calcined at 950 °C for 12 h and then at 1000 °C for 12 h, followed by quenched to room temperature naturally.

### 5.2.2 Characterization of the Structures and Properties

The XRD spectra were measured by D8 Advance diffractometer (Bruker, Germany) with Cu Kα radiation ($\lambda = 1.5406$ Å) source, operated at 40 kV and 40 mA. The spectra were collected at the 2θ range of 10–80°, with the step size of 0.02° and the counting time was 0.5 s. In situ XRD patterns were measured by an in situ stainless-steel cell, which was fabricated by using a beryllium window as the current collector to allow the penetration of the X-rays. The cathodes made of $Li_2RuO_3$ and B0.15-LRO were just covered by the beryllium window. A glass fiber (GF/D) made by Whatman was used as the separator and lithium metal as the anode. In situ XRD data were collected with the cell being charged and discharged in a current rate of 20 mA/g simultaneously. Rietveld refinement was performed via TOPAS software. XPS data were collected by AXIS Ultra instrument from Kratos Analytical (Al Kα radiation, $h\nu = 1486.6$ eV). The fitting of the spectra was performed by XPSPEAK41 software.

### 5.2.3 In Situ XAFS

In situ Ru K-edge XAFS spectra were collected at the beamline 1W1B of the Beijing Synchrotron Radiation Facility (BSRF) and the beamline BL14W1 of the Shanghai Synchrotron Radiation Facility and applying a Si (311) double-crystal monochromator. The in situ cells were fabricated on the basis of the cell used for the in situ XRD measurements that two beryllium windows were placed on both the cathode side and the anode side to allow the X-ray beam to penetrate. The spectra were collected when the cell was charged/discharged to a certain voltage at a slow current density of 30 mA $g^{-1}$. O K-edge and B K-edge XAS were collected from the beamline 4B9B of BSRF and the beamline BL10B of National Synchrotron Radiation Laboratory (NSRL). All data processing performed prior to analysis, including energy calibration, background removal, normalization, and Fourier transformation, was done using the software Athena.

### 5.2.4 *Preparation of the Electrodes and Electrochemical Tests*

The cathode electrodes were prepared via mixing 80 wt% active materials, 10 wt% polytetrafluoroethylene (PTFE) as the binder and 10% wt Super P as the conductive additive. And then, the mixture was pasted on an aluminum foil and rolling into a slice with a thickness of 25–30 μm. After that, the slice was cut into several small pieces as the individual cathode electrodes, followed by drying in a vacuum oven at 120 °C for at least 12 h before they were fabricated into the cells.

### 5.2.5 *First Principle Calculations*

DFT calculations were performed by the Vienna ab initio simulation package (VASP) using the projector augmented wave (PAW) approach. The spin-polarized generalized gradient approximation (GGA) was employed to investigate the structures, and the Perdew–Burke–Ernzerhof exchange–correlation functional was applied. The energy cutoff for the plane waves was set to 600 eV, and a Monkhorst–Pack scheme $4 \times 2 \times 2$ k-points mesh was used for integration in the irreducible Brillouin zone. The results were obtained within the GGA + U scheme with the Ru Hubbard U value of 5 eV.

## 5.3 Results and Discussions

### 5.3.1 *General Characterizations*

All of the samples were synthesized by sintering the mixture of $Li_2CO_3$, $RuO_2$, and $H_3BO_3$ in a proper proportion. To investigate the obtained products, XRD patterns were applied to study the crystal structure as shown in Fig. 5.1a. All of the patterns show perfect layered feature of C2/c space group without obvious impurity after the incorporation of certain amounts of boron. However, the (002) peak shows consecutive shift to lower angle, as reflected by the inset figure of Fig. 5.1a. This is probably

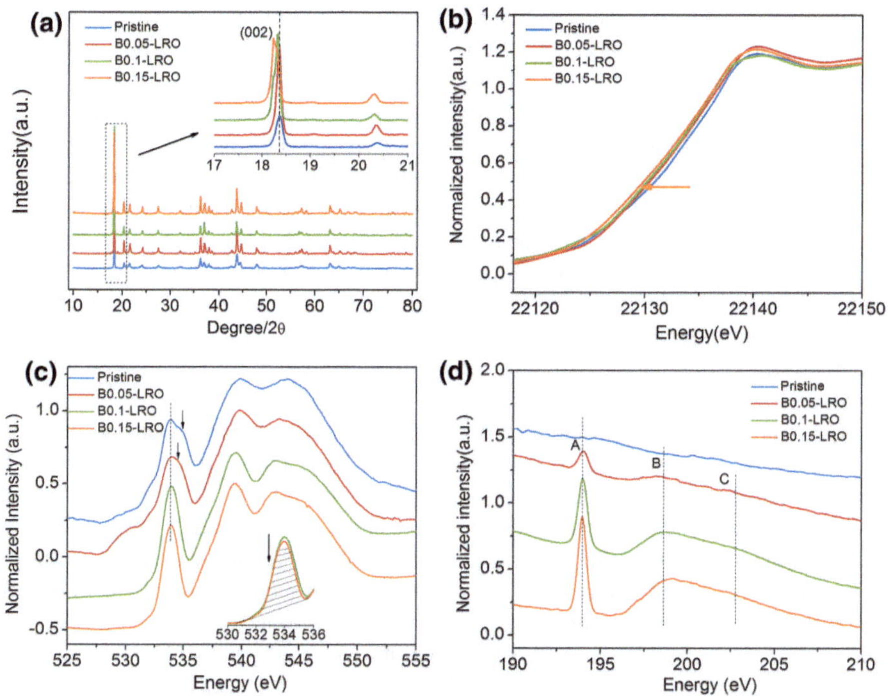


**Fig. 5.1** General characterization of $Li_2RuO_3$ and B*x*-LRO ($x = 0.05$, 0.1, and 0.15) samples. **a** XRD characterization. **b** Ru K-edge XAFS spectra. **c** O K-edge XAS spectra. **d** B K-edge XAS spectra

caused by doping boron atoms, which expands the space between the layers. Nevertheless, we did not find any change of the relative intensities of the peaks, suggesting that boron doping did not vary the long-range order. Due to the facilitation of boron to the calcination, the whole intensity of the diffraction peaks increases with the increase of the amounts of boron, indicating the degree of crystallinity is higher after boron's doping. The samples with higher boron content than B0.15-LRO were also prepared, however, with impurity like $RuO_2$ present. Therefore, the subsequent electronic and structural analysis of these impure samples is excluded.

The electronic structural variation of $Li_2RuO_3$ after the incorporation of boron atoms was studied. The variation of the electronic structure of Ru is characterized by Ru K-edge XAFS, as shown in Fig. 5.1b. The absorption edge of Ru has the tendency of shifting to a lower energy as the boron content increases, presenting a reasonable evolution as the electric neutrality requires lower valence of Ru to compensate for the charge brought by boron atoms. To investigate the electronic structural change of oxygen ions, we performed the soft X-ray absorption spectroscopy (XAS) technique in obtaining the O K-edge XAS spectra, as shown in Fig. 5.1c. The pre-edge peak of O K-edge XAS spectra corresponds to the transition of O 1s electrons to Ru(4d)-O(2p) hybridized states. As observed from Fig. 5.1c, the shape and intensity of the O K-edge spectra have large variations after different amounts of boron doping. The shoulder peak at about 535 eV for pristine sample attenuates for B0.05-LRO and then disappears for B0.1-LRO and B0.15-LRO. Due to the screening effect, the disappearing of the shoulder peak at higher energy indicates the increase of the electrons on the O 2p band or the decrease of the holes of O 2p band, which suggests the decrease of the covalence of Ru–O bond due to the incorporation of boron atoms. From B0.1-LRO to B0.15-LRO, the area of the normalized pre-edge peak decreases (see the inset of Fig. 5.1c), showing further decreases of electron holes on O 2p band and so does the covalence of Ru–O bond. All of these electronic structural changes provide potent evidence of the bulk-doping of the boron atoms after we excluded the possibility of Li- or Ru-deficiency in compensating the charge of boron by XPS shown below.

ICP is an advanced technique to probe the specific atomic concentration semi-quantitatively. However, the atomic mass of Li, Ru, and B atoms in $Li_2RuB_xO_3$ differs considerably (Li: 14, Ru: 101, B: 0.5–1.5), which makes it unable to obtain accurate atomic concentrations of the three elements simultaneously since ICP technique requires certain range of mass concentration of the elements in the solution to be measured. On the other hand, Li and B are light elements and hard to be full quantitatively determined, the real composition of B-doped samples is unlikely to obtain. We here provide the XPS semi-quantitative results of B0.15-LRO sample, which shows approximate composition (Li: Ru: B = 2.15:1:0.174) to the theoretical value (Li: Ru: B = 2:1:0.15). The results are shown in Fig. 5.2 and Table 5.1.

It is also very crucial to determine the actual site of doped boron atoms in the structure. The most probable case is that boron atoms reside in the pseudo-tetrahedral interstitial sites of $Li_2RuO_3$. Boron atoms are generally considered to be coordinated with three or four oxygen ions, and no evidence of $BO_6$ octahedron has been found, which suggests boron atoms are not likely to substitute Ru ions. Since the radius of $B^{3+}$ ions is only 0.027 nm, the interstitial-site doping is possible, as unraveled by previous works studying B-doped $TiO_2$ [23, 24]. Previous investigation on boron-doped $LiCoO_2$ and Li-rich manganese-based materials further proves that the doping of B in tetrahedral interstitial sites of layered rock-salt-type structure is possible [25, 26]. To probe the actual state of the doped boron atoms, B K-edge XAS spectra were collected as shown in Fig. 5.1d. The signal of boron becomes stronger as the content of boron increase. The sharp peak A at about 194 eV corresponds to trigonal $BO_3$ group while the broader peak B at 197−199 eV corresponds to $BO_4$ group, and the broad peak C is contributed by $BO_4$ group at 200–202 eV and $BO_3$ group at 203–204 eV jointly, as inferred from previous works [27–29]. We can determine that in the layered structure, boron atoms formed as both $BO_3$ and $BO_4$ groups and stayed as local polyanions. Furthermore, the Rietveld refinement analysis of the XRD patterns was performed to investigate the doping site. By utilizing $Li_2RuO_3$ model, the refinement converged to a result with Rwp = 6.49% (Fig. 5.3a). After introducing the boron atoms into the interstitial site, the refinement was improved with the Rwp = 6.02% (Fig. 5.3b). Although boron atoms have a small scattering cross section and

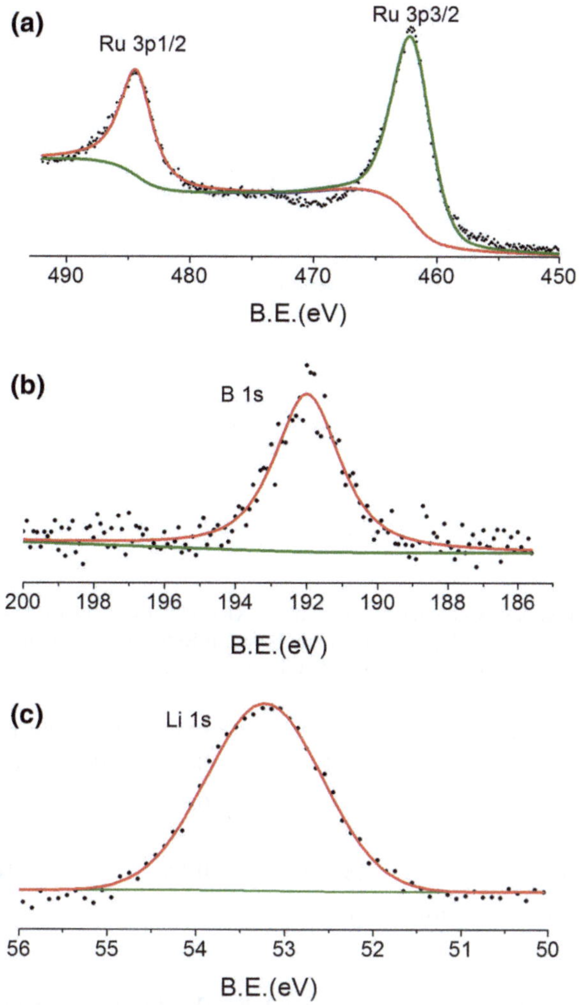


**Fig. 5.2** XPS spectra of **a** Ru 3p, **b** B 1s, and **c** Li 1s

influence the XRD reflections with limited extent, the improvement of the Rwp value still positively supports the existence of boron atoms in the interstitial sites of layered $Li_2RuO_3$.

### *5.3.2 Electrochemical Performances*

The electrochemical performance of $Li_2RuO_3$ and its boron-doped samples was tested via galvanostatic charge–discharge mode. The first and second charge and

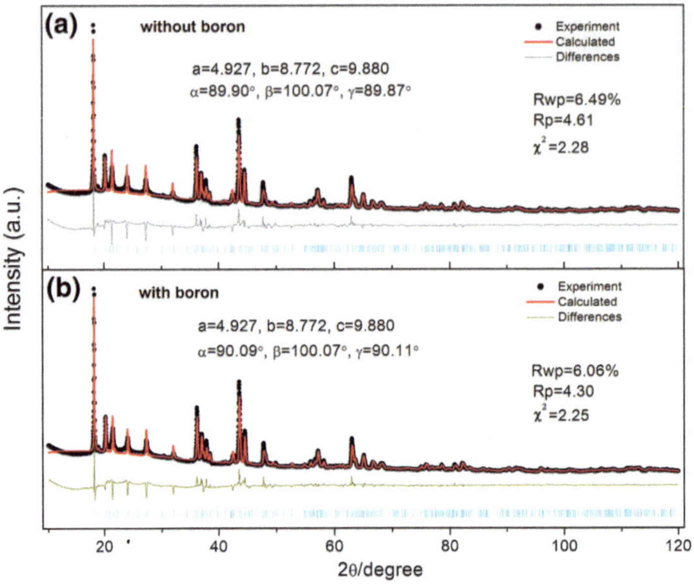


**Fig. 5.3** Rietveld refinement of the XRD patterns of the B0.15-LRO sample. **a** Refinement with the model of $Li_2RuO_3$ without any boron. **b** Refinement with the model of $Li_2RuB_{0.15}O_3$ with boron atoms reside in the pseudo-tetrahedral interstitial site

**Table 5.1** XPS semi-quantitative analysis of B0.15-LRO sample

| Peak | Area (I) | Sensitivity (S)[a] | Area/Senti. (I/S) | Atomic ratio |
|---|---|---|---|---|
| Li 1s | 1714.56 | 0.057 | 30080.00 | 2.147 |
| Ru 3p1/2 | 48200.69 | 3.44 | 14011.83 | 1 |
| B 1s | 1187.39 | 0.486 | 2443.19 | 0.174 |

[a] Scofield's sensitivity factors

discharge curves and derived dQ/dV plots are presented in Fig. 5.4. The first charge process of $Li_2RuO_3$ has been identified to consist of two processes involving Ru redox and oxygen redox previously, corresponding to the 3.6 and 4.2 V voltage platform, respectively, [10, 12, 30] as shown in Fig. 5.4a. However, after the doping of boron, the first charge behavior has been largely changed that the 3.6 V platform stretches while the platform of oxygen redox is shortened, as observed from the charge curves of B$x$-LRO ($x$ = 0.05, 0.1, 0.15) in Fig. 5.4a. This suggests that Ru takes more responsibility of the charge compensation process while oxygen takes less, though the total capacity decreases with the increase of the boron content. Other than that, the voltage of the oxygen redox plateau was enhanced from 4.2 V to higher voltages of 4.3–4.4 V after boron doping, as displayed by the dQ/dV plots in Fig. 5.4a. This enhancement is consistent with the inductive effect of polyanions proposed by Good enough in polyanion-type cathodes like $LiFePO_4$ [25]. It should be noted that though the plateaus are lifted, the redox processes of Ru and oxygen are totally completed before 4.6 V, not disrupted. This is crucial for the comparison between $Li_2RuO_3$ and Bx-LRO samples.

What interests us more is the voltage plateaus in the first charge process are of higher straightness than the pristine one, as reflected by the sharper peaks of the reduction and oxidation in dQ/dV plots. For the first discharge curves, the pristine sample shows a slope feature with a volcano-type peak displayed in the dQ/dV plot, while the doped samples present two independent plateaus and two distinctive needle-like reductive peaks in the dQ/dV plots. These differences on the charge/discharge behavior are more evident in the second charge and discharge processes, as shown in Fig. 5.4b. The charge and discharge curves both exhibit distinctive voltage plateaus and oxidative and reductive peaks for the boron-doped samples, compared with that of pristine $Li_2RuO_3$. All of these changes are supposed to be caused by boron doping, which induces the variation of the electronic structure and thus the electrochemical behavior. The relative size of the two peaks (labeled as "A" and "B") in the dQ/dV plots during the discharge of second cycle shown in Fig. 5.4b changes as a function of boron content, especially for the B0.15-LRO sample; specifically, the intensity of peak A increases relative to peak B as boron content increases. Since the peak intensity of dQ/dV plot is in direct relation to the slope of the voltage platform, the relative size change of the dQ/dV plots suggests the variation of the voltage platform, which may be resulted from the influences to the two-phase transition process caused by doping different amounts of boron. Specifically, the two-phase transition reaction corresponding to the peak A becomes more perfect with the increase of boron content than that corresponding to peak B at lower voltage.

More than that, the oxygen redox plateau at about 4.3 V is still observed for Bx-LRO samples in the second cycle, demonstrating a better structural retention and redox reversibility than pristine $Li_2RuO_3$. We also provide the charge/discharge performance at 15th cycle (Fig. 5.5), which demonstrates similar difference between the pristine and Bx-LRO samples. It is obvious that the electronic structural variation is responsible for the difference between the samples. More boron, lower the capacity, but higher the stability, as proved by the cycling performance shown in Fig. 5.6. For boron-doped samples, the capacity in the first cycle is lower than that of second cycle, which probably caused by the activation effect due to their larger particles than pristine sample. Though smaller particle size may enhance the cycling stability in most of the cases, such as Si anode, however, in our work, boron-doped samples with larger particles demonstrate better cycling ability than pristine $Li_2RuO_3$. This further stands out the effect of boron doping in enhancing the cycling stability. It is speculated that the first charge process of $Li_2RuO_3$ will make the total structure rearranged after the loss of amounts of oxygen, while it is inhibited by the tuning of the electronic structure with boron atoms. Only in such situation, the structure of boron-doped layered oxides can be reserved without any oxygen loss.

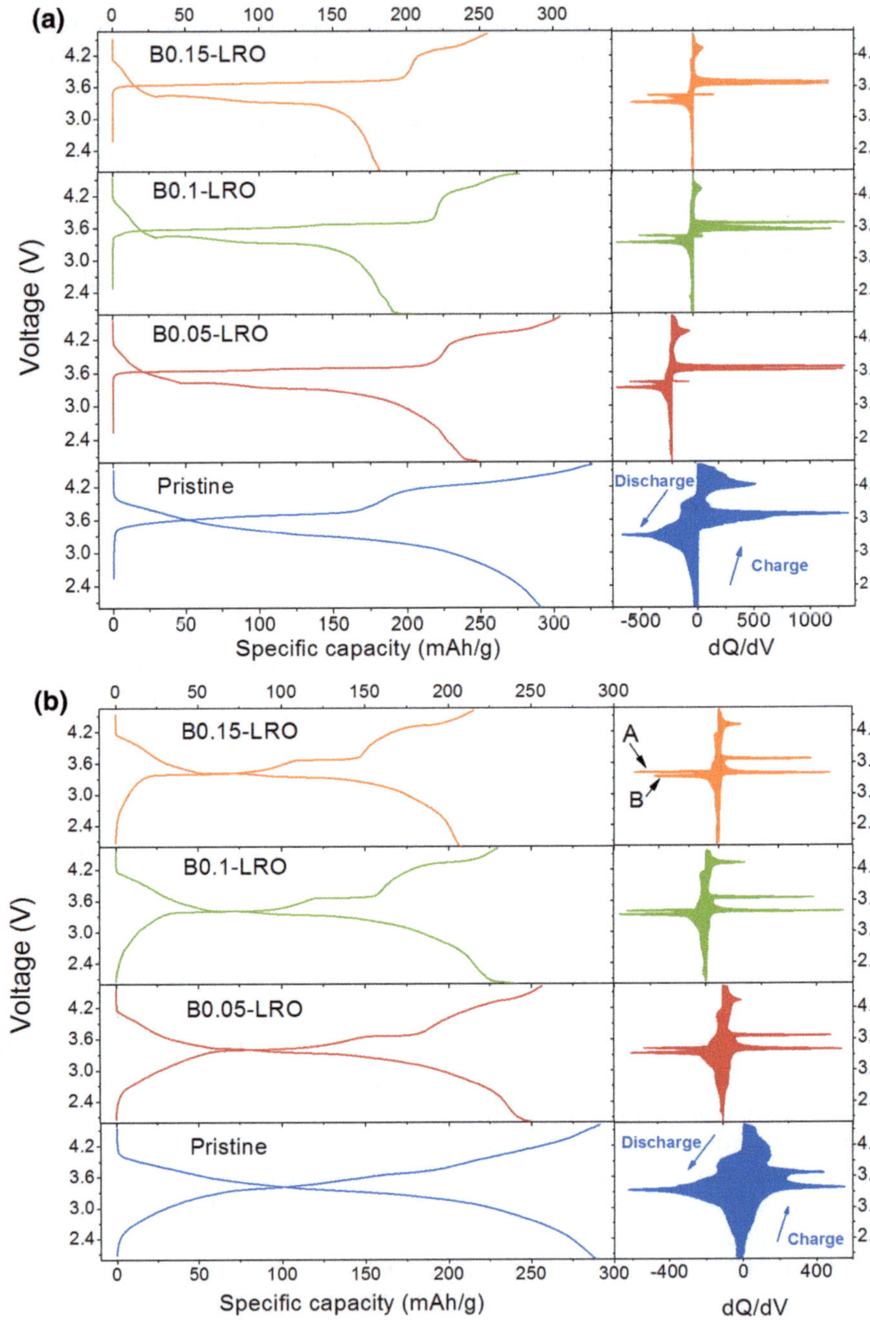


**Fig. 5.4** Electrochemical performance of $Li_2RuO_3$ and Bx-LRO ($x = 0.05$, 0.1, and 0.15). **a** The first charge and discharge curves and dQ/dV plots. **b** The second charge and discharge curves and dQ/dV plots

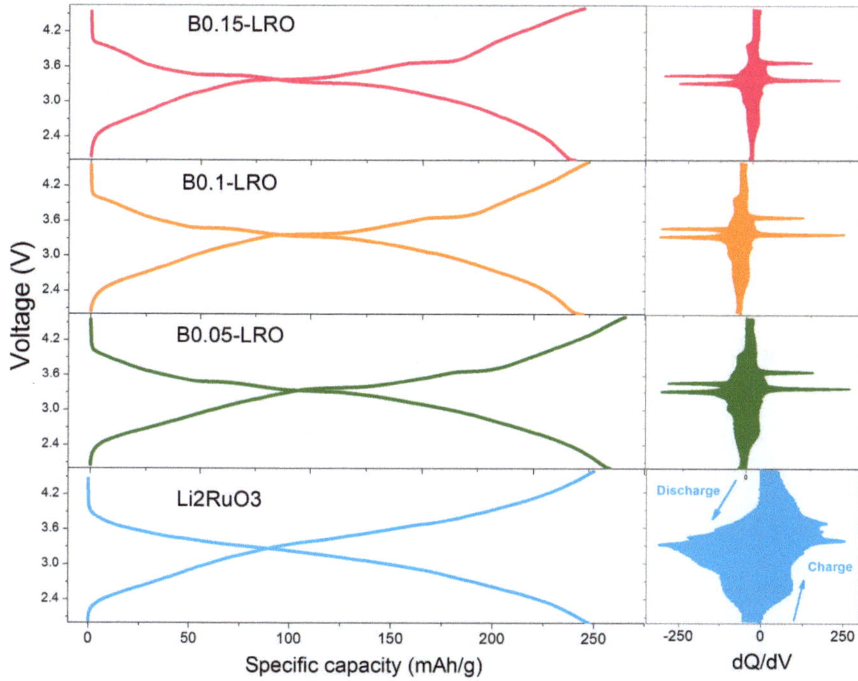


**Fig. 5.5** Charge and discharge curves and dQ/dV plots of $Li_2RuB_xO_3$ (x = 0, 0.05, 0.1, and 0.15)

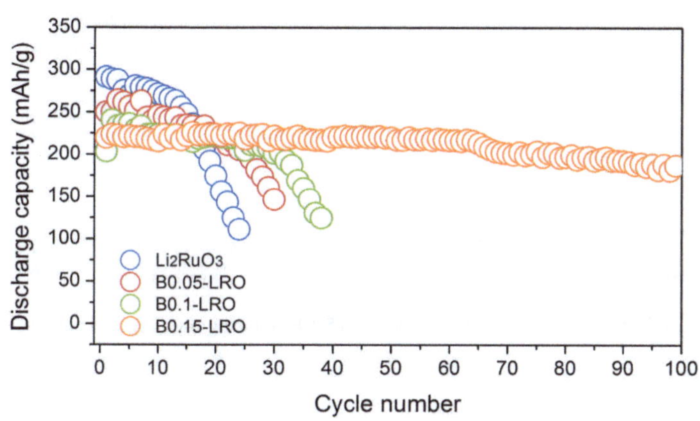


**Fig. 5.6** Cycling performance of the electrodes

### *5.3.3 In Situ XRD Study*

To verify our speculation, we took the in situ XRD to track the long-range-order evolution of the structure before and after boron doping. Figure 5.7 shows the in situ

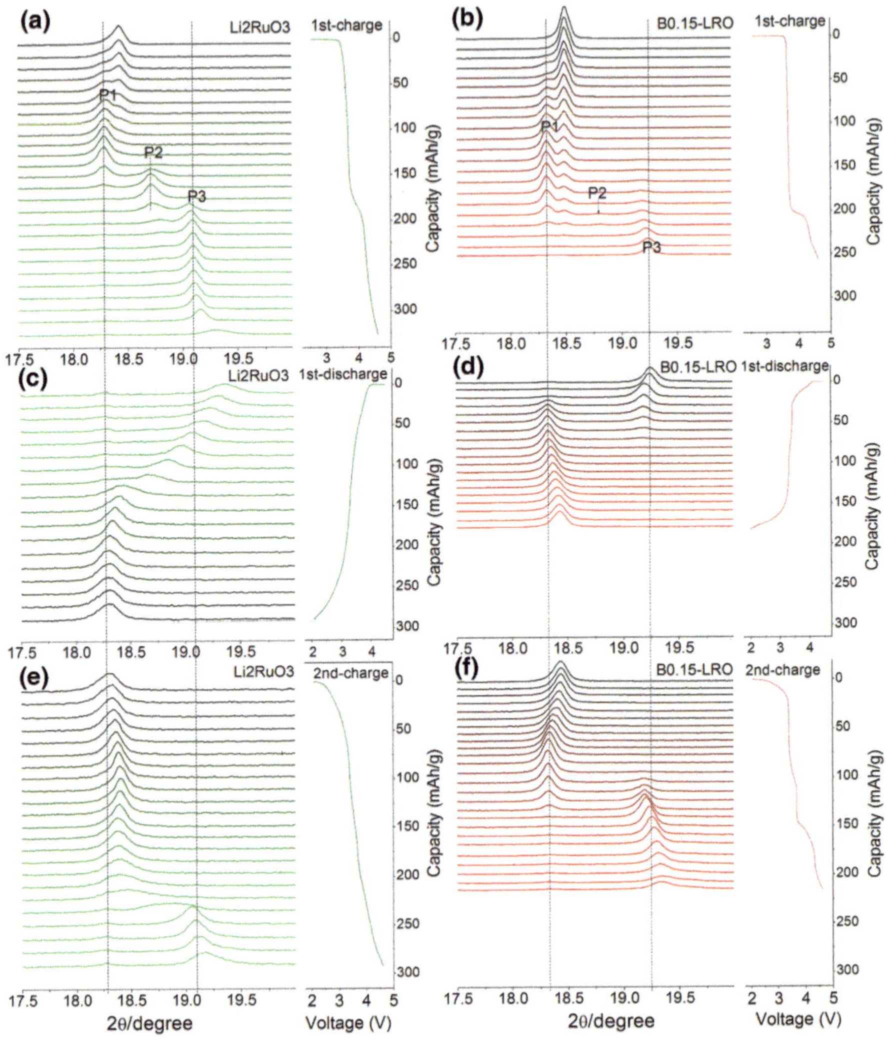


**Fig. 5.7** In situ XRD results of $Li_2RuO_3$ and B0.15-LRO during the first cycle and second charge process

XRD results of $Li_2RuO_3$ and B0.15-LRO between 17.5 and 20° for the first cycle and second charge process, respectively. The first charge process of $Li_2RuO_3$ shows a continuous three-phase-transition feature of the (002) peak in the first plateau as observed in Fig. 5.7a, with three interphases labeled as P1, P2, and P3 between $Li_2RuO_3$ and $RuO_3$. However, for B0.15-LRO, the phase transition process demonstrates much distinctive behavior as the P1 phase lasted for a longer process and the P2 phase appeared dependently in the process of P3 phase transition rather than an independent phase change presented in $Li_2RuO_3$, as shown in Fig. 5.7b. This difference may be induced by the variation of the electronic structure caused by the incorporation of boron. During the second plateau of the first charge process, a solid-solution structural evolution was presented for both $Li_2RuO_3$ and B0.15-LRO. However, the (002) peak of $Li_2RuO_3$ shifts for 0.25° while B0.15-LRO shows little change, suggesting the latter has less structural destruction by the oxygen redox process. Other than that, the broadening of (002) peak of $Li_2RuO_3$ demonstrates large internal stress existed in the crystal due to the induced structural distortion in oxygen redox process. There is no obvious broadening of the (002) peak in B0.15-LRO sample (Fig. 5.7b), demonstrating little internal stress and higher stability of the modified layered structure.

Figure 5.7c, d presents the in situ XRD results of the first discharge process of $Li_2RuO_3$ and B0.15-LRO, respectively. The evolution of the (002) peak of $Li_2RuO_3$ presents a two-stage feature: Before 3.3 V, the (002) peak takes continuous substantial shift during the oxygen redox process, demonstrating a solid-solution behavior; after that no shifting presents but the peak becomes broadening. However, for B0.15-LRO, the peak evolution shows a two-phase transition behavior rather than a solid-solution change before 3.3 V. The difference of structural evolution between the first charge process and the discharge process of $Li_2RuO_3$ is caused by the structural reorganization due to the loss of oxygen, resembling the situation of Li-rich manganese-based materials. Here, the B0.15-LRO remaining as a two-phase transition change in the discharge process can be the identity of better maintaining of the layered structure in oxygen redox process or reflect a better reversibility of the oxygen redox compared with $Li_2RuO_3$. Such difference also extends to the second charge process; see in Fig. 5.7e, f. By comparing the initial position of the (002) peak with that at the end of charge and discharge processes, we also can find that B0.15-LRO shows better reversibility than LRO. These results of in situ XRD are consistent with our previous speculation from the electrochemical performance and conclude that boron doping tuned the reversibility of the oxygen redox and thus the structure.

### *5.3.4 In Situ Ru K-Edge XAFS Study*

To further exploring the redox process of the total charge and discharge process before and after the doping, we performed the in situ Ru K-edge XAFS measurement, shown in Fig. 5.8. During the first charge process, the absorption edge of Ru K-edge shifts to higher energy before 4.2 V, but shifts to lower energy when charged to 4.6 V (Fig. 5.8a). This result is consistent with the previous work and was determined as the reductive coupling mechanism by Tarascon's group [10, 12]. Correspondingly, the intensity of the peaks of Fourier-transformed EXAFS spectrum, denoting the local structural disorder, shows similar evolution with the electronic structure (Fig. 5.8b). The two peaks of the Ru–O shell and Ru–Ru shell all demonstrate increased intensity before 4.2 V, while after that the intensity decreases drastically, indicating the local structure was disordered induced by the reductive coupling mechanism [30]. However, for the first charge process of B0.15-LRO, the Ru K-edge presents continuous

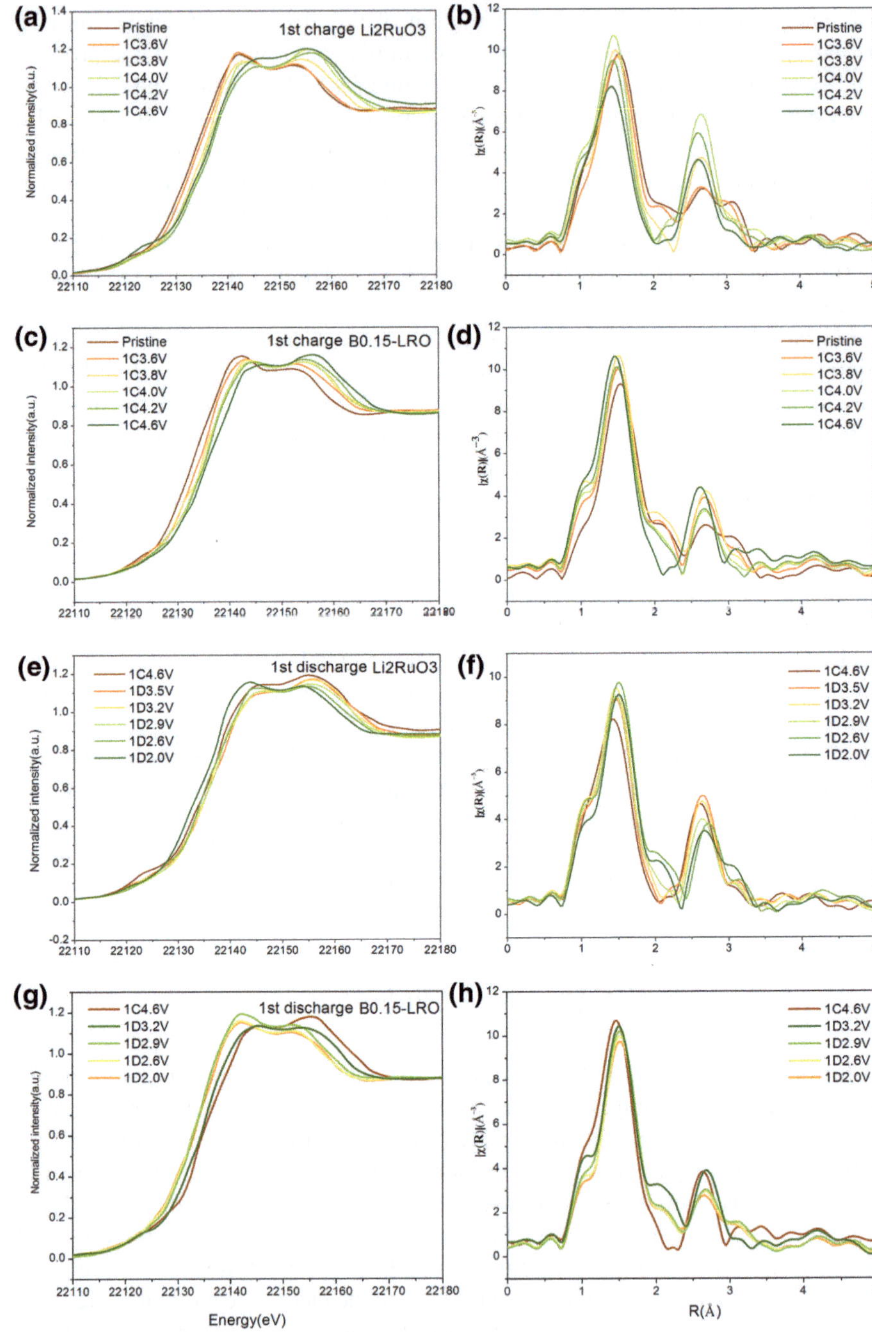


Fig. 5.8 In situ Ru K-edge XAFS spectra' results

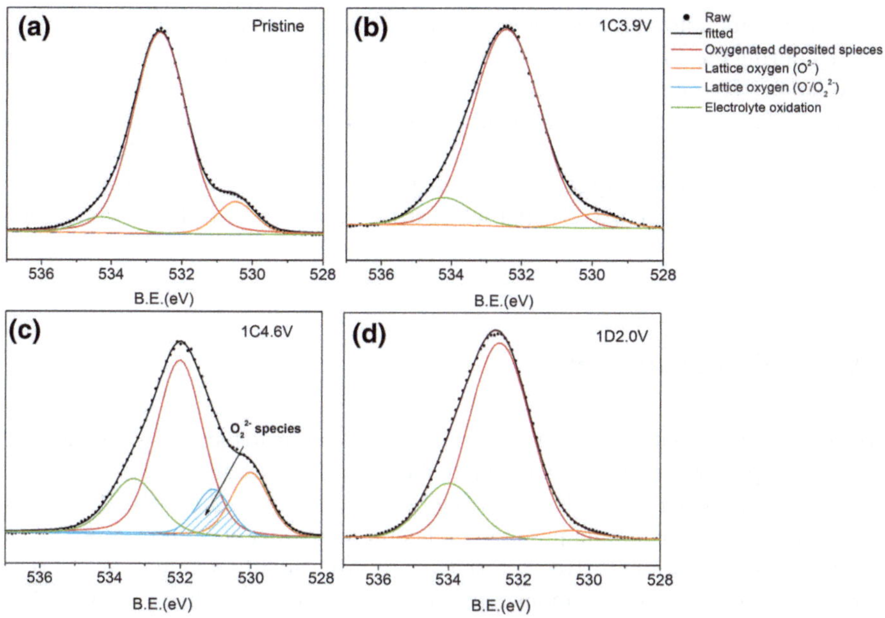


**Fig. 5.9** O 1s XPS spectra of B0.15-LRO at different charge/discharge states

shift to higher energy without any fluctuation, which is different from the case of $Li_2RuO_3$. Analogously, the evolution of the radial distribution functions (Fig. 5.8d) is generally monotonous during the whole process with little exceptions, suggesting that the local structure is becoming ordered during the total charge process, even though within the oxygen redox process. The electronic structural change of oxygen for B0.15-LRO was characterized by O 1s XPS analysis (Fig. 5.9), which exhibits a reversible oxygen redox during the charge and discharge processes.

Such difference between $Li_2RuO_3$ and B0.15-LRO is also determined in the discharge process, as seen in Fig. 5.8e, h. The valence of Ru ions in $Li_2RuO_3$ increases first when discharged from 4.6 to 3.5 V, and then decreases after that, while it is monotonous increase for B0.15-LRO. The intensity of the peaks in the radial distribution functions of $Li_2RuO_3$ shown in Fig. 5.8f tends to increase first and then decrease, while for B0.15-LRO the total process exhibits monotonous increase. This further demonstrates the absence of reductive coupling mechanism in B0.15-LRO though with the presence of oxygen redox.

As we have investigated before, the whole charge process of $Li_2RuO_3$ can be divided into two stages: In the lower voltage range called Stage L, Ru ions are oxidized with the structure more ordered; in the higher voltage range called Stage H, Ru ions are reduced successively and O ions are oxidized, with the structure more disordered. In Stage H, the reductive coupling mechanism happens when the oxygen is oxidized by a certain extent that the local structure will be distorted to relieve the energy. However, after the doping of boron, the reductive coupling mechanism was replaced by a monotonous electronic structural change, with the local structure more ordered during the charge process. Since the reduction of Ru ions begins at the start of oxidation of oxygen, the variation of the charge compensation mechanism should be contributed by the electronic structural change after the doping though the oxygen redox process was limited to smaller extent. The modified structure of Ru–O–B can make itself stable under oxygen redox without reductive coupling mechanism; thus, Ru ion is oxidized successively and the structure tends to be ordered without any distortion.

### *5.3.5 DFT Calculations*

In order to further uncover the intrinsic mechanism leading to the change of the redox process, we performed DFT calculation to theoretically manage with such question. The calculation was realized by a supercell of $Li_{16}Ru_8O_{24}$ and $Li_{16}Ru_8BO_{24}$, with the boron atom resides in the pseudo-tetrahedral interstitial site, and the optimized structure is shown in Fig. 5.10a, c, respectively. After totally delithiation, the optimized structure of $Ru_8O_{24}$ demonstrates large distortion as exhibited in Fig. 5.10b, with O–O dimers formed as the $O_2^{2-}$-like species and the length of O–O bond is 2.418 Å. The local view of optimized $Ru_8BO_{24}$ is shown in Fig. 5.10d, with less distortion compared with $Ru_8O_{24}$, and the average bond length of O–O bonds is about 2.515 Å, which is larger than that of $Ru_8O_{24}$. This implies that the oxidation of oxygen in B-doped $Li_2RuO_3$ is limited, and the structure exhibits higher order, which well confirms the experimental results. In order to probe the inherent mechanism, we plot the partial density of states (PDOS) to observe the electronic structural change before and after boron doping, as shown in Fig. 5.10e, f. The majority of the electrons residing around the Fermi level belong to Ru, and oxygen overlaps with Ru between the energy gaps of (−2, 0). Sequentially, the Ru ion will first give out its electrons upon delithiation, and then oxygen participates the redox reaction via the reductive coupling mechanism. After boron doping, the occupation of oxygen electrons between the energy gaps of (−2, 0) was reduced from 50.12 to 46.32%, suggesting the overlap between Ru 4d states with O 2p states is reduced. This result consists of previous electronic structural characterization that the covalence of Ru–O is reduced. As a result, Ru ions will take more responsibility on redox reaction, and oxygen redox will be limited to a certain extent. To figure out why reductive coupling mechanism is absent in boron-doped samples, we compared the PDOS before and after the structural optimization (demonstrated as rigid structure and relaxed, respectively, see Fig. 5.11). The results show that for half-delithiated $LiRuO_3$, the relaxed density of states (DOS) demonstrates no obvious change compared with rigid structure DOS, suggesting $LiRuO_3$ is a self-stable structure. For B-$LiRuO_3$, the rigid structure and relaxed DOS all present comparative lower energy of the oxygen 2p states than $LiRuO_3$, indicating a higher stability of the oxygen. When it comes to the totally delithiated $RuO_3$, the relaxed DOS demonstrates large variation compared with rigid structure DOS that the system changes from a conductor to a semicon-

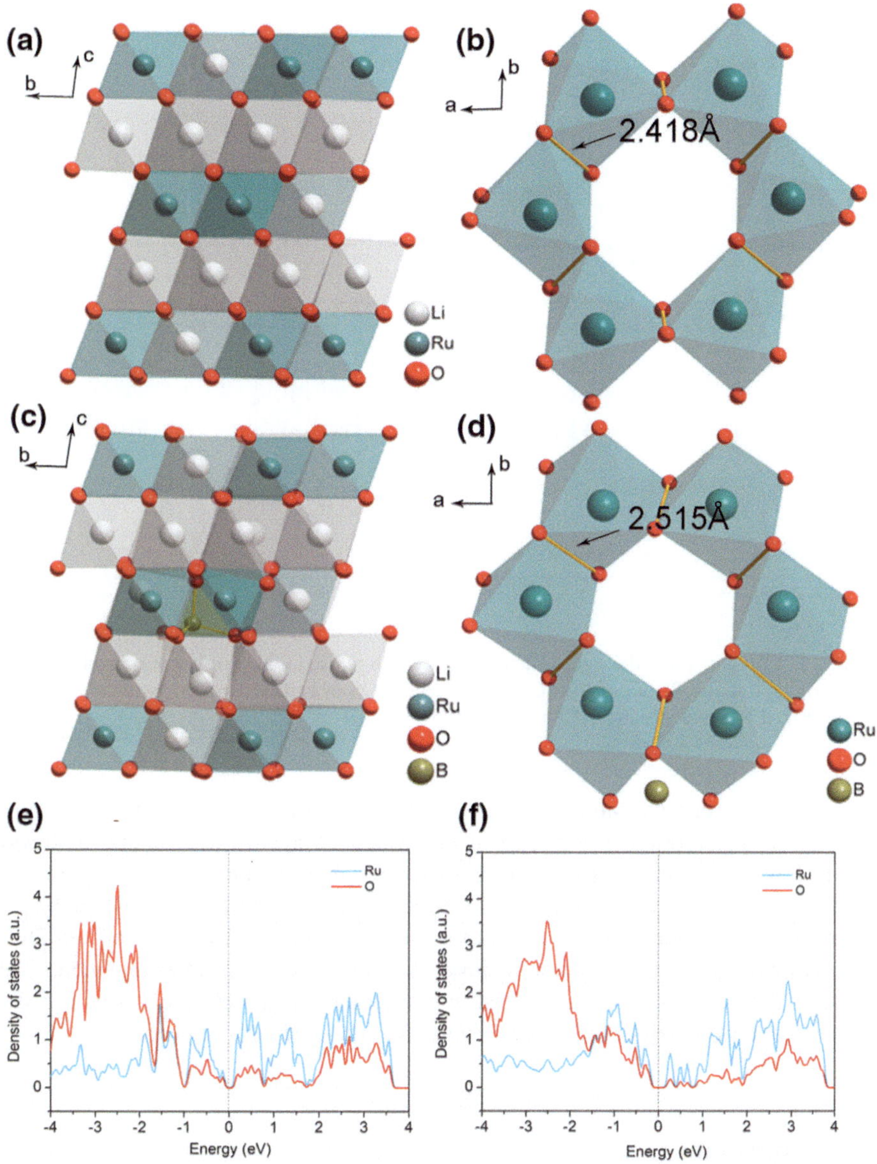


**Fig. 5.10** **a** Schematic of the optimized crystal model of Li16Ru8O24. **b** Local structure of the calculated Ru8O24 viewed from the c axis. **c** Schematic of the optimized crystal model of Li16Ru8BO24. **d** Local structure of the calculated Ru8O24 viewed from the c axis. **e** Calculated PDOS of Li16Ru8O24. **f** Calculated PDOS of Li16Ru8BO24

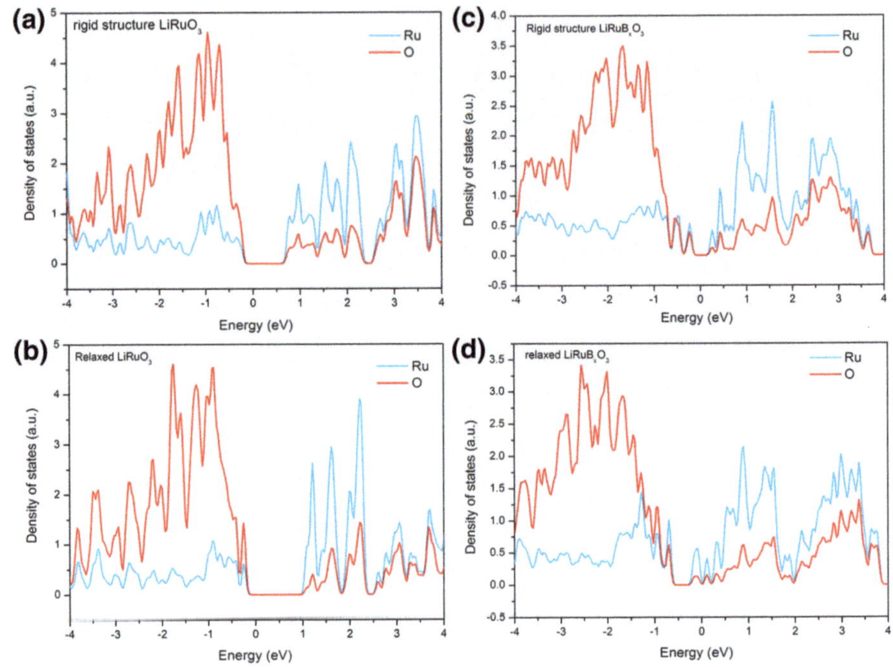


**Fig. 5.11** Comparison of the rigid structure PDOS with relaxed PDOS of $LiRuO_3$ and B-$LiRuO_3$. **a** The rigid structure PDOS of $LiRuO_3$. **b** Relaxed PDOS of $LiRuO_3$. **c** Rigid structure PDOS of B-$LiRuO_3$. **d** Relaxed structure PDOS of B-$LiRuO_3$

ductor with an energy gap (Fig. 5.12). Such electronic structural rearrangement is realized by the reductive coupling mechanism leading to the structure distortion and to lower the energy of oxygen 2p electrons, thus maintains the whole layered structure. However, the boron-doped sample after total delithiation (B-$RuO_3$) just shows an energy decrease for the comparison of rigid structure and relaxed DOS, still maintaining as a conductor without electronic structural rearrangement. This suggests that the B-$RuO_3$ is a stable structure rather than a metastable structure after a certain extent oxidation of oxygen due to the incorporation of boron.

## 5.4 Summary

The stability of the layered structure is related to the reversibility of the oxygen redox when the lithium-rich layered oxides deliver high capacity. Though with the reductive coupling mechanism, the excessive oxidation of the oxygen leads to a large capacity fading during cycles. After boron doping, the oxygen redox was tuned to a proper extent with the totally delithiated structure stable, without being distorted to relieve the energy. Upon that, the materials show better reversibility of the structural

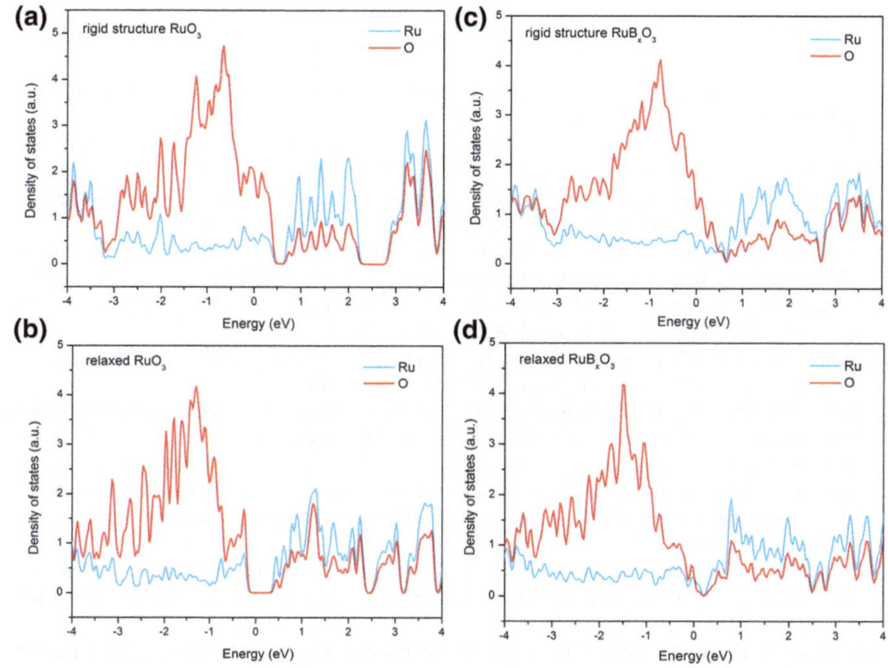


**Fig. 5.12** Comparison of the rigid structure PDOS with relaxed PDOS of $RuO_3$ and B-$RuO_3$. **a** The rigid structure PDOS of $RuO_3$. **b** Relaxed PDOS of $RuO_3$. **c** Rigid structure PDOS of B-$RuO_3$. **d** Relaxed structure PDOS of B-$RuO_3$

evolution during charge and discharge process, and deliver excellent cycling stability. The central appeal of this work is to provide a general strategy to balance the capacity and stability by tuning the reversibility of anionic redox in lithium-rich high-capacity electrodes. This balance means acceptable capacity as well as its stability during cycles. Simply lowering the cutoff voltage is an alternative strategy, but in some extent may not be as good as tuning its intrinsic electronic structure. We refer "tuning" since we just provide the strategy and corresponding experimental certifications, however, with not probably the best performance. In our work, B0.15 may not be the best choice to reach a balance; possibly, it is B0.12 or B0.13 to obtain higher capacities with acceptable stabilities. The choice of B0.05, B0.1, and B0.15 was just aimed to present a tendency of the electronic structural variation.

In general, through this work, we demonstrated that the oxygen redox can be tuned by modifying the electronic structure of metal–oxygen group and thus realized the balance between high capacity and high stability. For most of the lithium-rich layered oxides, the high capacities cannot be sustainable during long cycles due to the excessive oxidation of oxygen ions. This work will provide a scope for the modification of these lithium-rich layered systems to obtain stable high-capacity electrodes.

## References


1. Tarascon JM, Armand M. Issues and challenges facing rechargeable lithium batteries. Nature. 2001;414(6861):359–67.
2. Whittingham MS. Lithium batteries and cathode materials. Chem Rev. 2004;104(10):4271–302.
3. Ammundsen B, Paulsen J. Novel lithium-ion cathode materials based on layered manganese oxides. Adv Mater. 2001;13(12–13):943–56.
4. Thackeray MM, Johnson CS, Vaughey JT, et al. Advances in manganese-oxide 'composite' electrodes for lithium-ion batteries. J Mater Chem. 2005;15(23):2257.
5. Thackeray MM, Kang S-H, Johnson CS, et al. $Li_2MnO_3$-stabilized $LiMO_2$ (M = Mn, Ni, Co) electrodes for lithium-ion batteries. J Mater Chem. 2007;17(30):3112.
6. Bettge M, Li Y, Gallagher K, et al. Voltage fade of layered oxides: its measurement and impact on energy density. J Electrochem Soc. 2013;160(11):A2046–55.
7. Gallagher KG, Croy JR, Balasubramanian M, et al. Correlating hysteresis and voltage fade in lithium- and manganese-rich layered transition-metal oxide electrodes. Electrochem Commun. 2013;33:96–8.
8. Yabuuchi N, Takeuchi M, Nakayama M, et al. High-capacity electrode materials for rechargeable lithium batteries: $Li_3NbO_4$-based system with cation-disordered rocksalt structure. Proc Natl Acad Sci. 2015;112(25):7650–5.
9. Yabuuchi N, Takeuchi M, Komaba S, et al. Synthesis and electrochemical properties of $Li_{1.3}Nb_{0.3}V_{0.4}O_2$ as a positive electrode material for rechargeable lithium batteries. Chem Commun. 2016;52(10):2051–4.
10. Sathiya M, Rousse G, Ramesha K, et al. Reversible anionic redox chemistry in high-capacity layered-oxide electrodes. Nat Mater. 2013;12(9):827–35.
11. Sathiya M, Ramesha K, Rousse G, et al. High performance $Li_2Ru_{1-y}Mn_yO_3$ ($0.2 \leq y \leq 0.8$) cathode materials for rechargeable lithium-ion batteries: their understanding. Chem Mater. 2013;25(7):1121–31.
12. Sathiya M, Abakumov AM, Foix D, et al. Origin of voltage decay in high-capacity layered oxide electrodes. Nat Mater. 2015;14(2):230–8.
13. Ma J, Gao Y, Wang Z, et al. Structural and electrochemical stability of Li-rich layer structured $Li_2MoO_3$ in air. J Power Sources. 2014;258:314–20.
14. Ma J, Zhou Y-N, Gao Y, et al. Feasibility of using $Li_2MoO_3$ in constructing Li-rich high energy density cathode materials. Chem Mater. 2014;26(10):3256–62.
15. Lee J, Urban A, Li X, et al. Unlocking the potential of cation-disordered oxides for rechargeable lithium batteries. Science. 2014;343(6170):519–22.
16. Lee J, Seo D-H, Balasubramanian M, et al. A new class of high capacity cation-disordered oxides for rechargeable lithium batteries: Li–Ni–Ti–Mo oxides. Energy Environ Sci. 2015;8(11):3255–65.
17. Wang R, Li X, Liu L, et al. A disordered rock-salt Li-excess cathode material with high capacity and substantial oxygen redox activity: $Li_{1.25}Nb_{0.25}Mn_{0.5}O_2$. Electrochem Commun. 2015;60:70–3.
18. Saubanère M, McCalla E, Tarascon JM, et al. The intriguing question of anionic redox in high-energy density cathodes for Li-ion batteries. Energy Environ Sci. 2016;9(3):984–91.
19. Xie Y, Saubanère M, Doublet ML. Requirements for reversible extra-capacity in Li-rich layered oxides for Li-ion batteries. Energy Environ Sci. 2017;10(1):266–74.
20. Wolverton C, Zunger A. First-principles prediction of vacancy order-disorder and intercalation battery voltages in $Li_xCoO_2$. Phys Rev Lett. 1998;81(3):606–9.
21. Yoon WS, Kim KB, Kim MG, et al. Oxygen contribution on Li-ion intercalation – deintercalation in $LiCoO_2$ investigated by O K-edge and Co L-edge X-ray absorption spectroscopy. J Phys Chem B. 2002;106(10):2526–32.
22. Mizokawa T, Wakisaka Y, Sudayama T, et al. Role of oxygen holes in $Li_xCoO_2$ revealed by soft X-ray spectroscopy. Phys Rev Lett. 2013;111(5):056404.

23. Chen D, Yang D, Wang Q, et al. Effects of boron doping on photocatalytic activity and microstructure of titanium dioxide nanoparticles. Ind Eng Chem Res. 2006;45(12):4110–6.
24. Subramanian A, Wang H-W. Effects of boron doping in $TiO_2$ nanotubes and the performance of dye-sensitized solar cells. Appl Surf Sc.i. 2012;258(17):6479–84.
25. Li B, Yan H, Ma J, et al. Manipulating the electronic structure of Li-rich manganese-based oxide using polyanions: towards better electrochemical performance. Adv Func Mater. 2014;24(32):5112–8.
26. Alcántara R, Lavela P, Tirado JL, et al. Structure and electrochemical properties of boron-doped $LiCoO_2$. J Solid State Chem. 1997;134(2):265–73.
27. Fleet ME, Muthupari S. Coordination of boron in alkali borosilicate glasses using XANES. J Non-Cryst Solids. 1999;255(2–3):233–41.
28. Fleet ME, Muthupari S. Boron K-edge XANES of borate and borosilicate minerals. American Mineralogist. 2000;85(7–8):1009–21.
29. Liu X, Wang D, Liu G, et al. Distinct charge dynamics in battery electrodes revealed by in situ and operando soft X-ray spectroscopy. Nature Commun. 2013;4:2568.
30. Li B, Shao R, Yan H, et al. Understanding the stability for Li-rich layered oxide $Li_2RuO_3$ cathode. Adv Func Mater. 2016;26(9):1330–7.

# Chapter 6
# Thermodynamic Activation of Charge Transfer in Anionic Redox Process for Li-Ion Batteries

**Biao Li, Ning Jiang, Weifeng Huang, Huijun Yan, Yuxuan Zuo and Dingguo Xia**

## 6.1 Introduction

Rechargeable lithium-ion batteries with high energy densities are indispensable for the further development of portable electronics and electric vehicles. Traditional cathode electrodes based on materials such as $LiCoO_2$, $LiFePO_4$, and $LiMn_2O_4$ operate by reversible intercalation and extraction with restricted numbers of lithium ions during charge and discharge, respectively, resulting in limited specific capacities that cannot meet the demands of high-energy-density batteries [1–6]. However, the emergence of Li-rich Mn-based oxides ($Li[Li_xMn_yNi_zCo_{1-x-y-z}]O_2$), which exhibit the high capacities (beyond one lithium ion) favoured by anionic redox, promises exciting potential for future cathode development. Recently, new lithium-rich layered oxides, irrespective of structural order or disorder, were proposed based on the large contributions of oxygen to the redox process; these include $Li_2RuO_3^-$, $Li_3NbO_4^-$, $Li_2MoO_3^-$, and $Li_2IrO_3$-based oxides [7–16]. Although these materials present limited potential for practical use because of their expense, as investigational subjects they enrich our understanding of anionic redox processes and should facilitate the invention of usable high-capacity electrodes.

---

This chapter was adapted from B. Li et al. *Advanced Functional Materials* 2018, 28(4), 1704864 with permission. Copyright 2018, Wiley-VCH.

---

B. Li · N. Jiang · W. Huang · H. Yan · Y. Zuo · D. Xia (✉)
Beijing Key Laboratory of Theory and Technology for Advanced Batteries Materials, College of Engineering, Peking University, Beijing, P. R. China
e-mail: dgxia@pku.edu.cn

B. Li
e-mail: libiao2008@pku.edu.cn

 
B. Li, *Studies on Anionic Redox in Li-Rich Cathode Materials of Li-Ion Batteries*,
Springer Theses, https://doi.org/10.1007/978-981-13-2847-3_6

A more imperative and realistic issue is to identify when an anionic redox process can be activated to fulfill the need for high capacity in electrode design. Conventionally, the trigger for oxygen redox has been attributed to the high covalence between TM (transition metal) and oxygen, i.e., the high overlap of TM *d* states with O 2*p* states, as represented by the well-known $LiCoO_2$, which shows oxygen redox activity upon deep lithium removal [17–27]. This ambiguously defined principle, however, is violated by the case of $Li_2MnO_3$, in which the Mn–O bond is not highly covalent, yet still manifests considerable oxygen redox activity, though special treatment should be imposed, such as high temperature, nanosizing, slowly charging. A further puzzle extends to the oxygen redox activities of other cation-inert electrodes that resemble $Li_2MnO_3$, such as $Li_2TiO_3$, $Li_2ZrO_3$, and $Li_3NbO_4$, which do not exhibit electrochemical activity, even at extremely low current density. These unresolved problems prompted us to consider the underlying principle that guides the general activation mechanism of the anionic redox process.

## 6.2 Methods

### 6.2.1 *Synthesis of $Li_2TiO_3$ and $Li_{1.2}Ti_{0.4}Fe_{0.4}O_2$*

Layered $Li_2TiO_3$ was synthesized by sintering stoichiometric amounts of $TiO_2$ (Xilong, ≥99.0%) and $LiOH{\cdot}H_2O$ (Sinopharm, ≥95%) at 900 °C after homogenization by pestle milling. Cubic $Li_2TiO_3$ was fabricated via a hydrothermal method by dissolving 0.4 g $TiO_2$ and 1.2 g $LiOH{\cdot}H_2O$ in 30 mL deionized water, and then reacting at 180 °C for 12 h. $Li_{1.2}Ti_{0.4}Fe_{0.4}O_2$ samples were hydrothermally synthesized similar to the process employed for cubic $Li_2TiO_3$, using a Fe–Ti hydroxide precursor and $LiOH{\cdot}H_2O$ at 220°C for 12 h. The Fe–Ti precursor was prepared via a coprecipitation method by slowly adding a mixed solution of $Fe_2(SO_4)_3$ (Sinopharm, ≥99.0%) and $Ti(SO_4)_2$ (Sinopharm, ≥96%) into KOH (Sinopharm, ≥85%) solution. Layered $Li_2TiS_3$ was prepared by sintering a homogeneous mixture of stoichiometric $Li_2S$ (1% excess, J&K, 98%), titanium powder (J&K, 99.4%), and sulfur powder (Alfa, 99.5%), after sealing in a small quartz tube under vacuum, at 750 °C for 12 h. $Li_{1.2}Ti_{0.6}Co_{0.2}S_2$ was fabricated by the same procedure as for $Li_2TiS_3$, but with the addition of cobalt powder.

### 6.2.2 *Characterization of the Structures and Properties*

XRD patterns were collected using a Bruker D8 Advance diffractometer (Bruker, Germany) equipped with a Cu Kα radiation source (λ = 1.5406 Å) and operated at 40 kV and 40 mA. The $Li_{1.2}Ti_{0.4}Fe_{0.4}O_2$ spectra were recorded in the range of $2\theta$ = 10°–120°, with a step of 0.02° and counting time of 8 s. Rietveld refinements were performed using the TOPAS software from Bruker. The morphology of the $Li_{1.2}Ti_{0.4}Fe_{0.4}O_2$ sample was characterized by transmission electron microscopy (TEM, FEI, TECNAI F20). ICP measurement was performed with an inductively coupled plasma-atomic emission spectrometer (Prodigy 7, Leeman). All of the samples were dissolved in aqua regia under 80 °C totally, respectively, and the solution was diluted to 1–10 μg/ml before sent to test.

### 6.2.3 Preparation of the Electrodes and Electrochemical Tests

Each cathode electrode was prepared by mixing 80 wt% active material, 10 wt% Super P as the conducting medium, and 10 wt% polytetrafluoroethylene (PTFE) as the binder, followed by rolling the mixture into a piece, and slicing into disks fitting the size of CR2032 coin-type cell. The cathode electrodes were dried at 120 °C for at least 12 h before fabricating the testing cells. CR2032 coin-type cells were fabricated in an MBRAUN glove box (Germany) at moisture and oxygen levels below 0.1 ppm. The proprietary high-voltage electrolyte was purchased from the Beijing Institute of Chemical Reagents. Li metal was used as the anode and Celgard 2320 film as the separator. The galvanostatic charge/discharge tests were performed using a Neware tester (China) at room temperature. Cyclic voltammetry and GITT measurements were carried out by METEK electrochemical workstation (Princeton Applied Research).

### 6.2.4 XAS and XPS Methods

Ti, Fe, and Co $K$-edge XAFS spectra were collected in transmission mode at beamline 1W2B of the Beijing Synchrotron Radiation Facility (BSRF) with a Si(111) double-crystal monochromator. The samples to be measured were prepared by charging or discharging the electrodes to the corresponding voltages and transferred to the test station after sealing in 3 M tape for protection from atmospheric moisture and oxygen. Energy calibration was realized using the first inflection points of Ti and Fe foils as references (Ti $K$-edge = 4966 eV, Fe $K$-edge = 7112 eV). The spectra below the pre-edge region were fitted to a straight line, and the post-edge backgrounds were subtracted based on the cubic spline procedure. The normalized EXAFS spectra were converted from $E$ space to $k$ space weighted by $k$ and then converted to $R$ space.

The Ti and Fe $L$-edge and O $K$-edge XAS spectra were measured at beamline 4B7B at the BSRF, and the S $K$-edge XAS spectra were collected from beamline 4B7A at the same facility. All samples were carefully protected from oxygen and moisture in an argon-filled bag before transferral into the chamber for measurement.

The XPS spectra were obtained using an Axis Ultra instrument from Kratos Analytical (Al K$\alpha$ radiation, h$\nu$ = 1486.6 eV). All of the samples were washed by dimethyl carbonate (DMC) after being charged/discharged to certain voltages. For protection from the air, the samples were sealed in an argon-filled tube before they were sent for measurement. The binding energy was calibrated by the C 1*s* peak at 284.8 eV. The spectra were fitted using CasaXPS (Casa Software Ltd.) software based on the least squares method. The peak fitting was applied with 70% Gaussian and 30% Lorentzian line shapes, and the full width at half maximum was kept in the range of 1–2.3 eV.

### 6.2.5 First Principle Calculations

The spin-polarized DFT + $U$ ($U$ = 5.6 eV for Fe) calculations were performed based on the Vienna ab initio simulation package (VASP), using a generalized gradient approximation (GGA) to investigate the structures and Perdew–Burke–Ernzerhof (PBE) functional to describe electron exchange and correlation. A plane-wave basis set with a cutoff of 600 eV was employed. A Monkhorst–Pack *k*-point mesh of $3 \times 1 \times 7$ was used for integration in the irreducible Brillouin zone for the energy optimization and DOS calculations. The supercell employed in the calculation was $Li_{36}Ti_{12}Fe_{12}O_{60}$, with Li, Ti, and Fe ions randomly distributed in the bulk. All lattice parameters and atom positions were fully relaxed with forces reaching less than 0.05 eV Å$^{-1}$. For the calculation of the polaron migration barriers in $Li_2TiO_3$, a supercell (221) containing 32 formula units (192 atoms) was used for C2/c, and a supercell containing 16 formula units (96 atoms) was employed for the Fm$\bar{3}$m structure. The DFT + $U$ method was used, and the $U$ values for Ti and O were 3.0 and 11.0 eV, respectively, in order to trap the electron holes. A Gamma-cantered $1 \times 1 \times 1$ *k*-point grid was used in the supercell calculations, and the *k*-point effect on polaron migration barriers was investigated. The error was less than 30 meV.

## 6.3 Results and Discussions

### 6.3.1 Structures and Properties of Layered and Cubic Rock-Salt $Li_2TiO_3$

In our investigation, we specifically selected $Li_2TiO_3$ as a model electrode material. Layered $Li_2TiO_3$ with C2/c symmetry shows no substantial capacity between 2.0–4.8 V at 10 and 1 mA g$^{-1}$, which may be explained by its electronic insulating nature (as proven by its 3.2 eV band gap, Fig. 6.1). However, disordered cubic Fm$\bar{3}$m $Li_2TiO_3$, which can be classified as a semiconductor since its calculated band gap is only 0.5 eV, still delivers no capacity, even at the extremely low

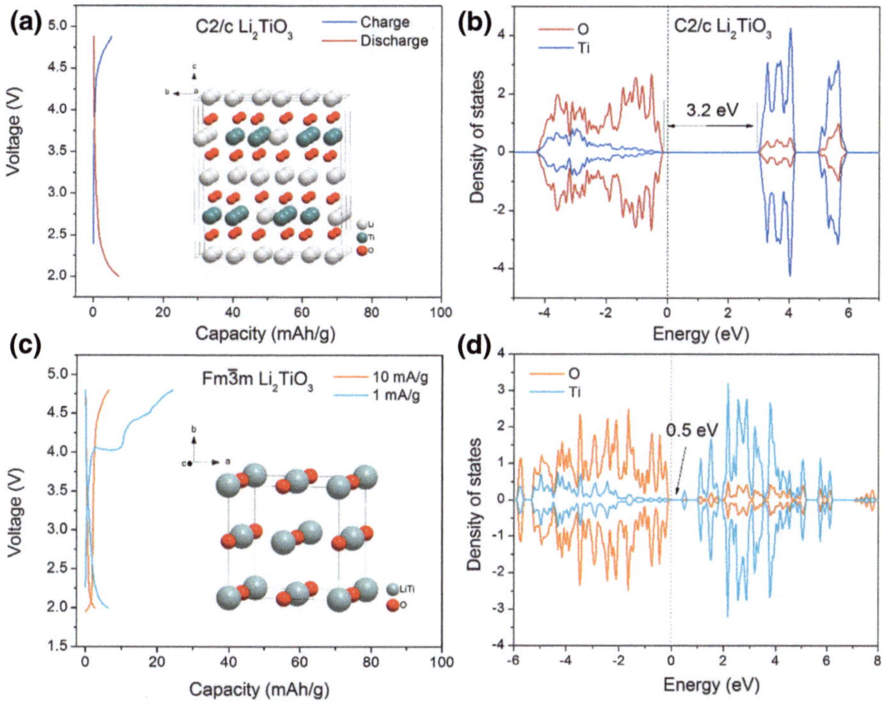


**Fig. 6.1** **a** Charge/discharge curves of C2/c $Li_2TiO_3$ at 10 and 1 mA $g^{-1}$. **b** The calculated PDOS of *C2/c* $Li_2TiO_3$. **c** Charge/discharge curves of Fm$\bar{3}$m $Li_2TiO_3$ at 10 and 1 mA $g^{-1}$. Though the charge process at 1 mA $g^{-1}$ shows a slight voltage plateau, it likely originates from defects or traces of active species whose contribution would be magnified by the extremely low current density, since the discharge process is irreversible. **d** Calculated PDOS of Fm$\bar{3}$m $Li_2TiO_3$

rate of 1 mA $g^{-1}$ (Fig. 6.1). Comparing the calculated band gaps of several common electrodes (Table 6.1) reveals that the band gap of $Li_2TiO_3$ is comparable to that of $LiCoO_2$ and even smaller than that of electrochemically active $LiCoPO_4$ and $LiMnPO_4$. Hence, it may not be feasible to explain the inactivity simply by the band gap of the cation inactive material. Another possibility for the inactivity is that the delithiation voltage of $Li_2TiO_3$ exceeds the voltage limits (>4.8 V). However, the theoretical predictions of delithiation voltages for both layered and cubic $Li_2TiO_3$ are below 4.8 V, and the latter has an even lower average voltage of 4.13 V for the extraction of the first 0.5 lithium ions (Fig. 6.2). So, the question remains as to the ultimate origin of the inactivity of $Li_2TiO_3$. It should be mentioned here that, though with particle size of only 31 nm, Fm$\bar{3}$m $Li_2TiO_3$ still has no activity under small current, which demonstrates its distinction with $Li_2MnO_3$, since the latter can be largely activated by nanosizing with tremendous capacity. Although the *d* orbitals of $Ti^{4+}$, $Zr^{4+}$, and $Sn^{4+}$ have no electrons, the anion *p* electrons from the ligand can transfer via the *p*–*d* transition to activate the anionic redox, as demonstrated previously by the electrochemical activity of $Li_2TiS_3$ [32].

**Table 6.1** Comparison of theoretically predicted band gaps

| Electrode | Calculated band gap (eV) | Refs. |
|---|---|---|
| $LiCoO_2$ | 2.7 | [28] |
| $LiFePO_4$ | 2.0 | [29] |
| $LiMnPO_4$ | 4.1 | [29] |
| $LiCoPO_4$ | 4.2 | [29] |
| $Li_2MnO_3$ | 1.8 | [30] |
| $Li_2SnO_3$ | 3.4 | Calculated in this work |
| $LiAlO_2$ | 5.4 | [31] |

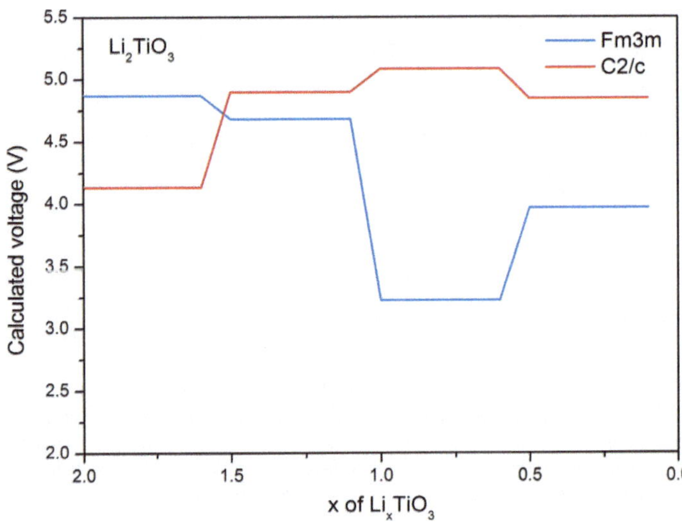


**Fig. 6.2** Predicted delithiation voltages of $Li_2TiO_3$ with different symmetries by DFT calculations

## 6.3.2 *General Properties and Electrochemical Performance of $Li_{1.2}Ti_{0.4}Fe_{0.4}O_2$ Electrodes*

To address this issue, we incorporated Fe ions into cubic $Li_2TiO_3$ (considering its lower band gap than layered $Li_2TiO_3$). The fabricated $Li_{1.2}Ti_{0.4}Fe_{0.4}O_2$ adopts mainly $Fm\bar{3}m$ symmetry, as shown by the X-ray diffraction (XRD) result in Fig. 6.3a.

The XRD pattern was fitted via the Rietveld method into two phases, $Fm\bar{3}m$ and C2/c, with corresponding phase fractions of 87.08 and 12.92%, respectively. The crystal structure of the main phase with $Fm\bar{3}m$ symmetry is shown in Fig. 6.3a (left inset figure) and presents cubic features with Fe ions randomly occupying positions originally held by Ti ions. However, the cubic structure of $Li_{1.2}Ti_{0.4}Fe_{0.4}O_2$ can also be viewed from a layered perspective (middle inset figure in Fig. 6.3a), which shows disordered cation distribution in the metal layers. Actually, the cubic and layered

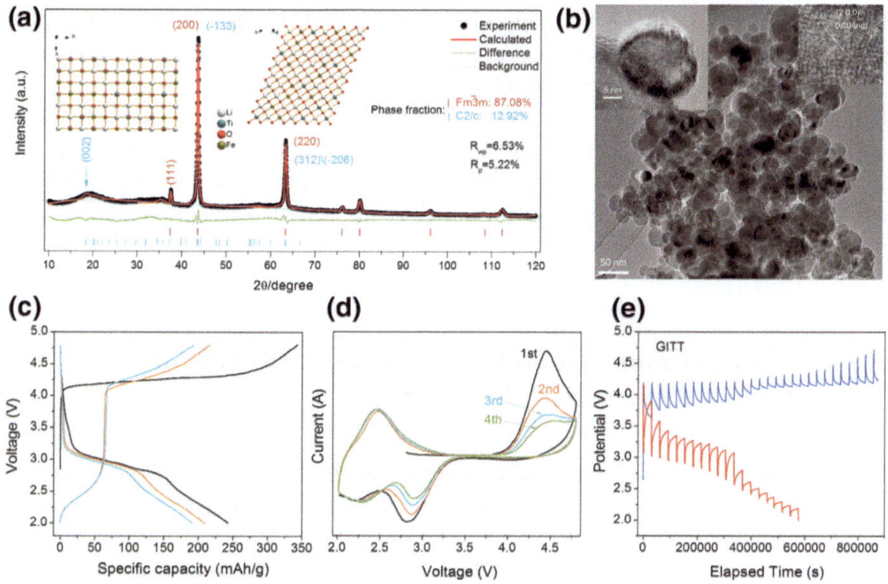


**Fig. 6.3** **a** Rietveld refined XRD pattern of $Li_{1.2}Ti_{0.4}Fe_{0.4}O_2$. **b** TEM image of the as-prepared $Li_{1.2}Ti_{0.4}Fe_{0.4}O_2$ sample. **c** Charge/discharge curves of the $Li_{1.2}Ti_{0.4}Fe_{0.4}O_2$ electrode in the range 2.0–4.8 V. **d** Cyclic voltammetry curves of $Li_{1.2}Ti_{0.4}Fe_{0.4}O_2$, with a scan rate of 0.1 mV $S^{-1}$. **e** GITT curve for $Li_{1.2}Ti_{0.4}Fe_{0.4}O_2$

phases are only discriminated by the ordering of the cations, as both phases adopt the same oxygen framework. A slight ordering of the lithium ions with respect to the TM ions generates the weak symmetry of the layered structure, as observed from the broad, weak (002) peak at 18.4° of the C2/c phase. The high-resolution transmission electron microscopy (TEM) image in Fig. 6.3b shows that $Li_{1.2}Ti_{0.4}Fe_{0.4}O_2$ crystallizes in circular nanoplates with an average diameter of 20 nm and presents the perfect crystal lattice of the (200) crystal face of the $Fm\bar{3}m$ structure, with a 0.204 nm spacing (inset figure of Fig. 6.3b).

$Li_{1.2}Ti_{0.4}Fe_{0.4}O_2$ delivers approximately 350 mAh $g^{-1}$ capacity in the first charge, and 250 mAh $g^{-1}$ in the discharge (Fig. 6.3c). The activation of the electrochemical activity by Fe substitution in $Li_2TiO_3$ is interesting and merits further study with regard to its redox behavior. Assuming the availability of $Fe^{4+/3+}$ can only account for 0.4 Li ions per formula, corresponding to a capacity of 130 mAh $g^{-1}$, which cannot explain the long voltage plateau in the first charge. Since the $Ti^{4+}$ species cannot be further oxidized, oxygen remains the only possibility by which to explain the charge compensation, which resembles the case of $Li_2MnO_3$. The subsequent discharge is characterized by two reduction peaks that are distinctly situated at about 2.8 and 2.3 V, while the second charge gives two well-separated oxidation peaks at 2.5 and 4.4 V (Fig. 6.3d). In view of the low potential, the peaks at 2.3 V (reduction) and 2.5 V (oxidation) are likely contributed by $Fe^{3+/2+}$, since peaks for $Ti^{4+/3+}$ normally appear below 2 V. The 4.4 and 2.8 V peaks, which exhibit large hysteresis, may correspond to a reversible oxygen redox process. This speculation was further confirmed by the galvanostatic intermittent titration technique (GITT) results shown in Fig. 6.3e. The first charge plateau presents a high overpotential up to 0.5 V, and the same feature is observed in the 2.8 V discharge plateau. This large voltage hysteresis is likely caused by the sluggish kinetics of the anionic redox process. For discharge after 2.8 V, the overpotential decreases enormously to 0.15 V, suggesting a transition of the redox mechanism.

### 6.3.3 Charge Compensation Mechanism of $Li_{1.2}Ti_{0.4}Fe_{0.4}O_2$ During Charge and Discharge

To further unravel the specific redox process, we systematically studied the electronic structural changes of Fe, Ti, and O via X-ray absorption spectroscopy (XAS) and X-ray photoelectron spectroscopy (XPS) techniques. Figure 6.4 shows the electronic structural variations of Ti and Fe in $Li_{1.2}Ti_{0.4}Fe_{0.4}O_2$ during the first cycle, investigated by the ex situ XAS technique. The specifically studied points of the first cycle are marked clearly in Fig. 6.4a. The absorption edges of the Ti $K$-edge XAFS spectra clearly show no rigid shift during the process, except for shape variations (Fig. 6.4b). This is expected since $Ti^{4+}$ ions cannot be further oxidized, although their local environments are disturbed upon lithium removal, as indicated by the regular variation of the Ti–O shell in the Fourier-transformed EXAFS results (Fig. 6.4b, inset figure). Fe ions exist as $Fe^{3+}$ (Fig. 6.4e) and have no valence change in the first cycle before discharge at 2.8 V, as suggested by the Fe $K$-edge XAFS spectra (Fig. 6.4c). Correspondingly, the Fourier-transformed EXAFS results (Fig. 6.4c, inset figure) reveal little variation in the Fe–O coordination. However, the absorption edge shifts rigidly to lower energy at 1D2.0V, accompanied by a sudden decrease in the Fe–O coordination peak intensity, demonstrating that further lithium insertion is charge-compensated by $Fe^{3+/2+}$. This shows great consistency with the reduction redox peak at 2.3 V (Fig. 6.3d) and the remarkably decreased overpotential inferred from the GITT result (Fig. 6.3e).

Additionally, the soft XAS technique was applied to track the electronic structural changes of Ti and Fe (Fig. 6.4d, e). It should be noted that the M–$L$ edge (M = Ti and Fe), corresponding to the electronic transition of $2p$ electrons to $3d$ states, is more sensitive to electronic structural change than the $K$ edge. Figure 6.4d clearly shows that the Ti–$L$ edge undergoes no shift during the first cycle, and the integrated area of the representative $L_3$-$t_{2g}$ peak changes little throughout the process (Fig. 6.4f). This demonstrates that the Ti $3d$ states remain unchanged during the entire charge/discharge process, in accordance with preceding analysis. As for the Fe $L$-edge XAS spectra, the peak position remains constant before 1D2.8V but shifts to lower energy at 1D2.0V (Fig. 6.4e), accompanied by a decrease in the $L_3$ peak intensity (Fig. 6.4f), which firmly proves the reduction of $Fe^{3+}$ to $Fe^{2+}$. In these

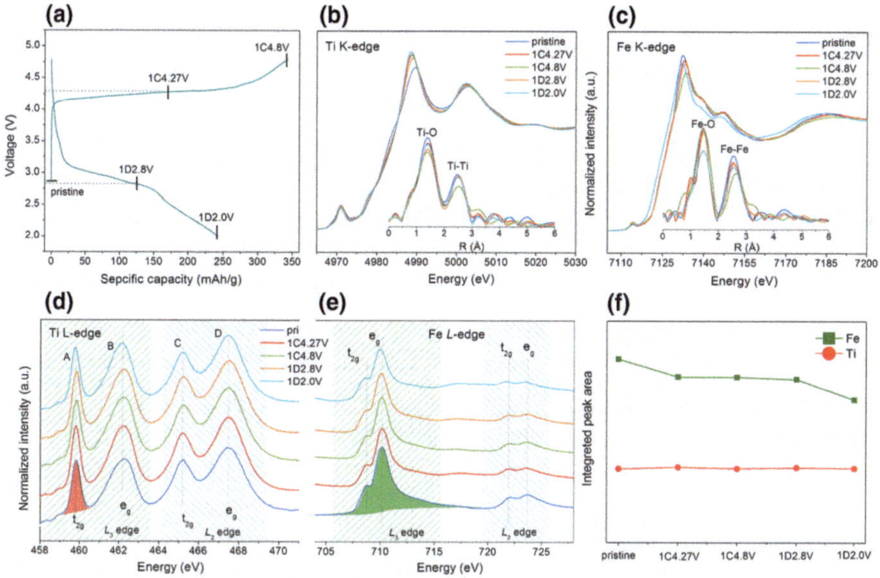


**Fig. 6.4** **a** Charge and discharge states in the first cycle chosen to study the electronic structural changes of Fe and Ti in $Li_{1.2}Ti_{0.4}Fe_{0.4}O_2$. **b** Ti *K*-edge XAFS spectra during the first cycle. The inset shows the Fourier-transformed $k^3$-weighted EXAFS spectra. **c** Fe *K*-edge XAFS spectra during the first cycle. The inset shows the Fourier-transformed $k^3$-weighted EXAFS spectra. **d** Ti $L_{3,2}$-edge XAS spectra. **e** Fe *L*-edge XAS spectra. Similar to the Ti *L*-edge, the Fe *L*-edge also splits into $L_3$ and $L_2$, and, further, to $t_{2g}$ and $e_g$, respectively. **f** Evolution of the integrated areas of the Ti $L_3$-$t_{2g}$ and Fe $L_3$-edge peaks in the respective soft XAS spectra

studies, a peculiar observation was made: The peak intensity of the Fe $L_3$-edge is also reduced during the charge of the pristine material to 1C4.27V.

Since Fe and Ti are not responsible for the charge compensation in either the first charge or the pre-stage discharge process before 2.8 V, the possibility of an oxygen redox process was studied via O 1*s* XPS and XAS techniques (Fig. 6.5). Figure 6.5a–e shows the O 1*s* XPS spectra fitted into several species in different charge/discharge states. From the pristine sample to 1C4.27V (Fig. 6.5a–c), a peak near 531.6 eV, corresponding to $O_2^{(2-n)-}$ species, emerges and grows more significant (the area proportion increases from 20.62 to 29.54%; see Fig. 6.6). Thus, the principle redox process in the first charge is dominated by $O^{2-}/O_2^{(2-n)-}$. From 1C4.8V to 1D2.8V, the peak from the $O_2^{(2-n)-}$ species decreases from 29.54 to 16.94%, suggesting the first plateau in the discharge curve is a reversible oxygen redox process. With further discharge to 1D2.0V, the $O_2^{(2-n)-}$ species persists with a nearly unchanged area (16.95%, Fig. 6.6) since the $Fe^{3+/2+}$ redox overtakes the process, in accordance with the prior analysis of TM ions. It should also be noted that the absolute changes of carbonate species and deposited species are meaningless, since the samples were washed by dimethyl carbonate to ensure the cleanness of the surface.

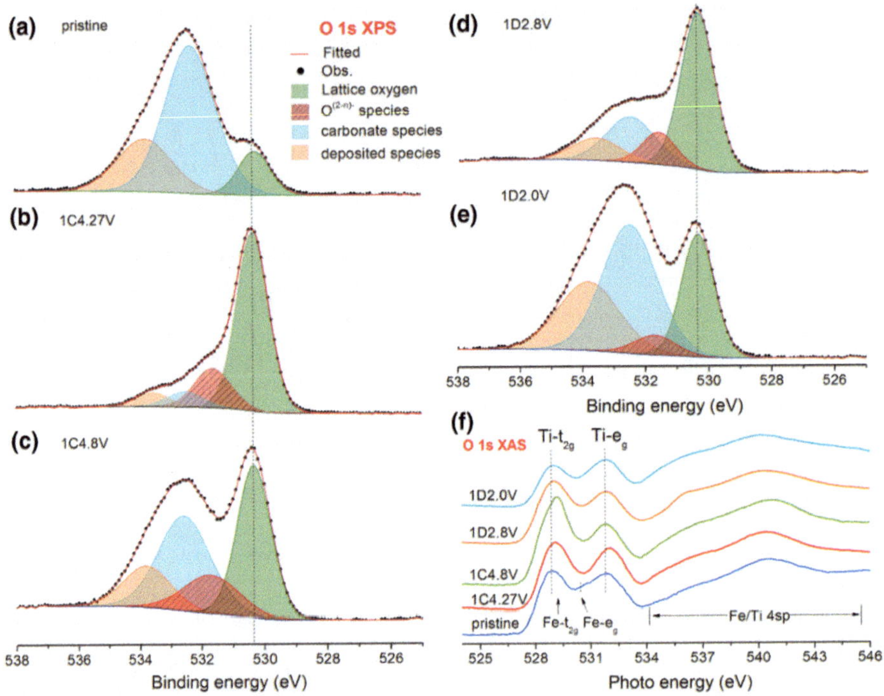


**Fig. 6.5** **a–e** O 1*s* XPS results for the different charge/discharge states. All spectra were deconvoluted into component species, as indicated by the shaded areas. **f** O 1*s* XAS spectra during the first cycle

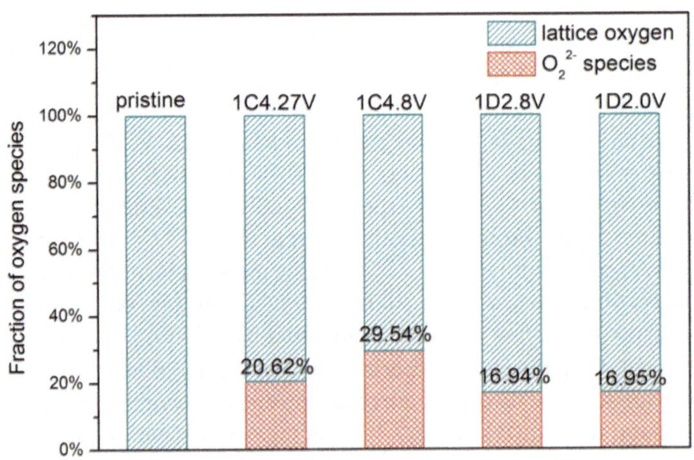


**Fig. 6.6** Fraction of oxygen species calculated from the fitted peak areas of the O 1*s* XPS spectra

We employed O 1*s* XAS spectra to obtain more specific information on the oxygen redox process (Fig. 6.5f). Since the average particle size is around 20 nm, the O 1*s* XAS spectra with probing depth of less than 10 nm can provide quasi-bulk information (the radius of the particle is around 10 nm; further, the particles are nanoplates with small thickness; see TEM figure in Fig. 6.3b). The pre-edge peak of the O 1*s* XAS spectrum originates from an electronic transition from the O 1*s* state to the unoccupied O 2*p* states mixed with M 3*d* states. The dissection of the individual peak was accomplished according to the actual electronic configuration (Fig. 6.7). The transitions from the O 1*s* state to the unoccupied Ti 3*d* states mixed with O 2*p* states predominate the pre-edge area due to the wholly empty 3*d* band, differentiated as the two main peaks corresponding to Ti $t_{2g}$ and $e_g$ bands (Fig. 6.5f). The transitions to the Fe $t_{2g}$ and $e_g$ bands (indicated by arrows in Fig. 6.5f) are nearly obscured by the strong Ti signals. To extract the information for the unoccupied O 2*p* states, the analysis was based on the metal *L*-edge results described above. Initially, we found that the Fe $e_g$ signal diminishes after the pristine electrode is charged to 4.27 V, implying a reduction in the number of d electron holes. This is well consistent with the Fe *L*-edge analysis, which also presented a decrease in the peak intensity (Fig. 6.4f). It seems that $Fe^{3+}$ is initially slightly reduced during the first charge, possibly through charge transfer, resembling the reductive coupling mechanism proposed by the Tarascon group [33]. Since there is no substantial variation in the Ti 3*d* band throughout the charge/discharge process, the increase in the Ti-$t_{2g}$ peak intensity relative to the Ti-$e_g$ peak in charging to 4.27 and 4.8 V (Fig. 6.5f) must be contributed principally by oxygen oxidation, which produces more unoccupied O 2*p* states. When discharged to 2.8 V, the Ti-$t_{2g}$ peak is weakened relative to the Ti-$e_g$ peak due to electron back-filling of the O 2*p* vacant states, revealing a reversible oxygen redox process. However, at the end of discharge, the Ti-$t_{2g}$ peak is further decreased relative to the Ti-$e_g$ peak, even compared with the pristine sample. This can be explained by the reduction of $Fe^{3+}$ to $Fe^{2+}$, which would largely diminish the Fe 3*d* electron holes, rather than the further reduction of oxygen, as supported by the XPS results. Hence, the O 1*s* XAS spectral results firmly prove the existence of a reversible oxygen redox process for $Li_{1.2}Ti_{0.4}Fe_{0.4}O_2$.

By this analysis, the large overpotential exhibited in the charge/discharge curves, especially for the anionic process, can be well understood. The high hysteresis is originated from both kinetic and thermodynamic origins; the former relates to the sluggish kinetics of the anionic redox process, as observed by the GITT test shown in Fig. 6.3e, while the latter may be caused by some thermodynamic reactions, e.g., reversible forming/breaking of the O–O bonds, reversible TM ions migration, which are all highly pertinent to anionic redox. Actually, partially reversible transitional metal migration is well established in LMLOs to explain the voltage decay and hysteresis issues, and further theoretical studies were performed on the $Li_2MnO_3$ model that attributes the origin of the migration to anionic redox. In addition, anionic redox-facilitated TM ion migration was also found in $Li_{2-x}Ru_{1-y}Ti_yO_2$ electrodes, which causes severe voltage decay. All these studies suggest that anionic redox will induce the formation of O–O dimers that opens the space for TM ions migration or lowers the migration energy barrier along the path. More work should be dedicated to

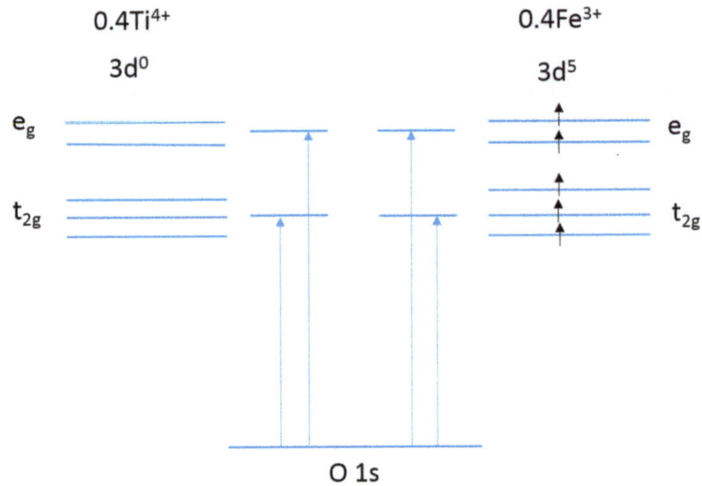


**Fig. 6.7** Analysis of the TM 3*d* contribution to the O 1*s* XAS pre-edge peak

figure out the inherent mechanisms, which are the next projects that we are focusing on but beyond the purpose of this work.

The remarkable capacity that $Li_{1.2}Ti_{0.4}Fe_{0.4}O_2$ provides via an oxygen redox process in the first charge without a contribution from the active metals led us to consider the role of Fe. We employed density functional theory (DFT) to investigate the charge compensation mechanism, as well as to interpret the absence of a Fe redox process. The calculated partial density of states (PDOSs) in Fig. 6.8a reveals an utterly abnormal electronic structure in which the O 2*p* states dominate in the vicinity of the Fermi level rather than the Fe 3*d* states; moreover, the latter reside in a very low energy region entirely below the O 2*p* states. We further plotted the Fe 3*d* electron-spin DOS (Fig. 6.8b), which shows a high-spin configuration for $3d^5$ ($t_{2g}^3 e_g^2$). This finding is experimentally validated by the profile of the Fe *L*-edge spectra according to previous reports [16]. We speculated that the high-spin state of $Fe^{3+}$ lowers the energy of the Fe 3*d* electrons and, thus, the whole system; fixing $Fe^{3+}$ in the low-spin state largely raises the energy, as demonstrated in Fig. 6.9. Therefore, with this unique electronic structure, only the O 2*p* electrons hop beyond the Fermi level to participate in the redox process, whereas Fe is inactive, as seen from the PDOS evolution during delithiation (Fig. 6.10).

Charge-density difference maps were also calculated to visualize the geometrical distribution of the redox centers (see Fig. 6.8c, d). Electron depletion occurs mainly at oxygen sites throughout the process, especially on pure non-bonding oxygen 2*p* bands directed toward Li rather than those overlapped with Ti and Fe *d* orbitals. This conforms to previous studies in which pure non-bonding O 2*p* states generated along the Li–O–Li direction have higher energy and are more easily involved in redox processes than the bonding states [34].

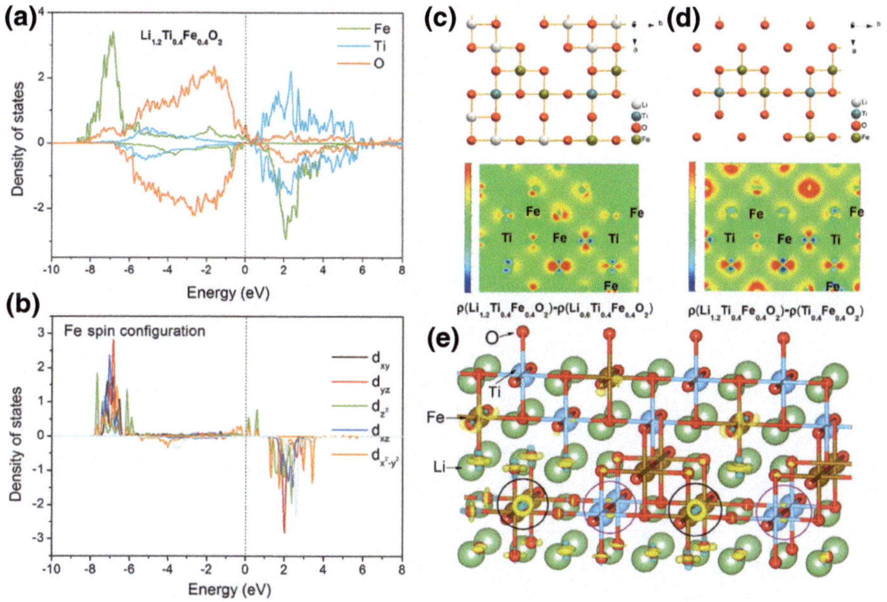


**Fig. 6.8** **a** Calculated PDOS of $Li_{1.2}Ti_{0.4}Fe_{0.4}O_2$. **b** Spin configuration of $Fe^{3+}$ ions in $Li_{1.2}Ti_{0.4}Fe_{0.4}O_2$. **c** Charge-density difference maps of $\rho(Li_{1.2}Ti_{0.4}Fe_{0.4}O_2) - \rho(Li_{0.6}Ti_{0.4}Fe_{0.4}O_2)$ and **d** $\rho(Li_{1.2}Ti_{0.4}Fe_{0.4}O_2) - \rho(Ti_{0.4}Fe_{0.4}O_2)$. **e** 3D view of the charge-density difference map of $\rho(Li_{1.2}Ti_{0.4}Fe_{0.4}O_2) - \rho(Ti_{0.4}Fe_{0.4}O_2)$

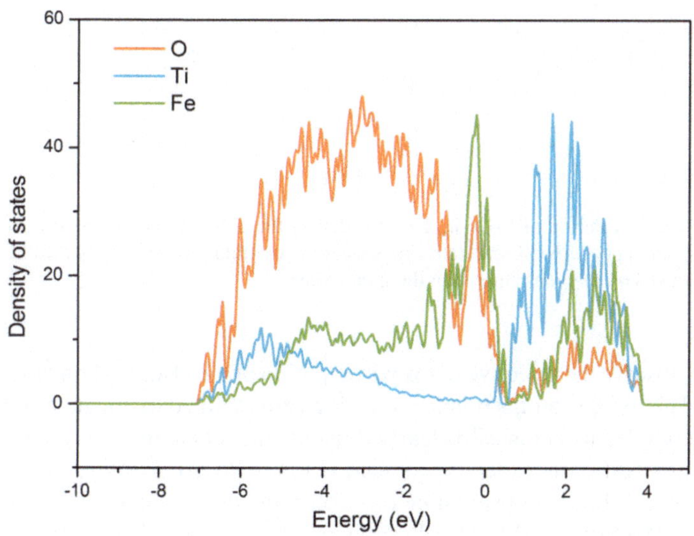


**Fig. 6.9** Calculated PDOS of $Li_{1.2}Ti_{0.4}Fe_{0.4}O_2$ with low-spin $Fe^{3+}$

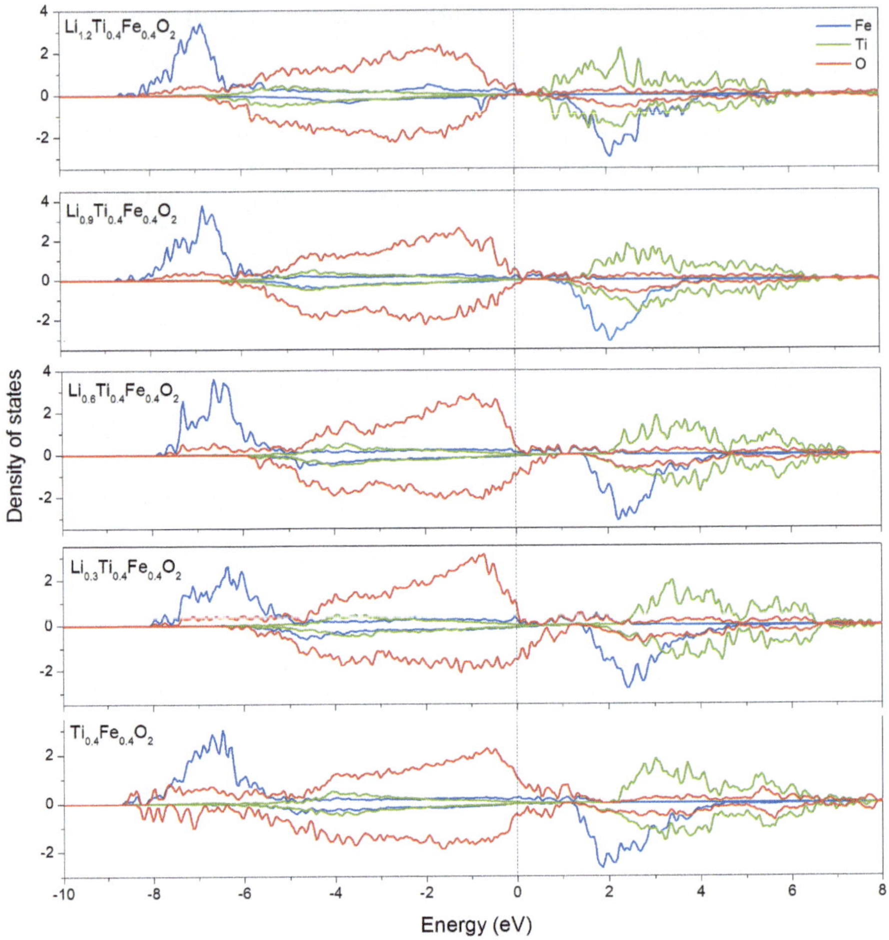


**Fig. 6.10** PDOS evolution of $Li_{1.2}Ti_{0.4}Fe_{0.4}O_2$ during the delithiation process obtained via DFT calculations. The unoccupied O 2*p* states increase upon delithiation, indicating the charge transfer is mainly focused on oxygen rather than the metal ions

We can observe the selective redox participation of the O atoms bonded to different TM ions. The oxygen atoms bonded to Fe ions are inclined to donate more electrons than those bonded to Ti ions. To clearly illustrate this phenomenon, a 3D view of the charge-density difference map is provided in Fig. 6.8e. Amazingly, almost all of the oxygen ions exhibiting considerable electron removal are bonded to Fe ions rather than Ti ions. We particularly chose the oxygen ions surrounded by only Fe ions or only Ti ions (here, Li ions are ignored due to their insignificant bonding effect) to observe the phenomena, as indicated by black and purple circles in Fig. 6.8e, respectively. Thus, the oxygen atoms bonded only to Fe ions are obviously electron depleted, whereas those bonded only to Ti ions exhibit little change. Furthermore, oxygen atoms bonded to both Fe and Ti ions present smaller electron density changes than those bound only to Fe ions. This selective redox participation of oxygen is peculiar, as we calculated that the O 2*p* orbitals have similar energy distributions for O ions coordinated only to Fe or only to Ti ions (Fig. 6.11).

This finding is very different from the case of $Li_2MnO_3$–$LiMO_2$ (M = Ni, Co, Mn, or a combination thereof) layered oxides, which present electron holes on the oxygen ions surrounded by $Mn^{4+}/Li^+$ ions. In the present case, $Li_2TiO_3$ forms the base of $Li_{1.2}Ti_{0.4}Fe_{0.4}O_2$, similar to $Li_2MnO_3$, but electron holes are generated on oxygen ions coordinated by Fe rather than Ti. This may be the origin of the difference between the electrochemical activities of $Li_2TiO_3$ (inactive) and $Li_2MnO_3$ (active). Alternatively, the Fe ions substituted into the $Li_2TiO_3$ lattice may facilitate electron transfer during the redox process and thus enhance the oxygen redox activity. Notably, although there is no change in the electronic structure of $Ti^{4+}$, $Fe^{3+}$ undergoes partial reduction during the initial charge as a result of electron transfer between Fe and O. These complex results suggest that charge transfer may only occur between O and certain transition metal ions such as Mn and Fe, and is inhibited for other metal ions like Ti and Nb. This ultimately prompts us to reconsider what determines the activation of an anionic redox process which would normally have been attributed to the high covalence of the metal–oxygen bonds.

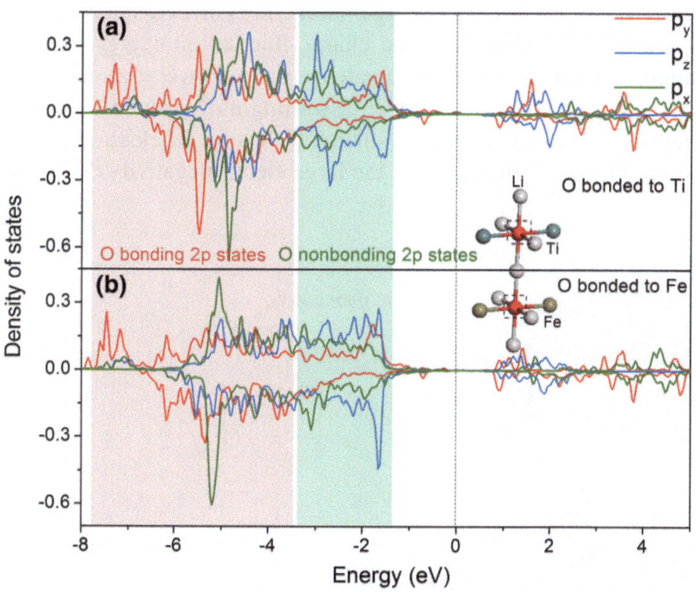


**Fig. 6.11** Isolated oxygen DOS analysis. **a** The DOS of an oxygen ion bonded to two Ti ions and four Li ions, as illustrated by the local structure view in the middle of the figure. **b** The DOS of an oxygen ion bonded to two Fe ions and four Li ions

### 6.3.4 Zaanen–Sawatzky–Allen Model in the Anionic Redox of $Li_{1.2}Ti_{0.4}Fe_{0.4}O_2$

To ascertain the basic principles underlying anionic redox activation, we considered the microscopic electron-transfer process during redox reactions in terms of solid-state physics theory. Generally, one electron is transferred to the current collector as one lithium ion is extracted. This transfer process is determined by two dimensionalities: thermodynamic dimension and kinetic dimension (Fig. 6.12). The kinetic process is normally strongly related to the conductivity of the electrode, determined by the energy barrier ($E_b$) of polaron (electron or hole) hopping between two atomic sites. $E_b$ does influence electrochemical performance kinetically; for example, the low conductivity in $LiFePO_4$ restrains its electrochemical activity at higher current rates. In this study, the $E_b$ values calculated for C2/c- and Fm$\overline{3}$m-$Li_2TiO_3$ (and compared with other electrodes in Fig. 6.13) demonstrate that the inactiveness of $Li_2TiO_3$ should not be attributed to its insulating properties. Therefore, the activity inhibition of $Li_2TiO_3$ may be of thermodynamic origin, which we shall next consider.

The redox mechanism is strongly associated with the formation of the first ionization or affinity state of the electrode upon charge/discharge, which can be traced to the relative values of the Mott–Hubbard $U$ and charge-transfer energy $\Delta$ originating from the Zaanen–Sawatzky–Allen model [35, 36]. The term $U$ is used to describe the $d$–$d$ Coulomb and exchange interactions involved in charge fluctuations of $d_i^n d_j^n \leftrightarrow d_i^{n-1} d_j^{n+1}$, while $\Delta$ is the charge-fluctuation energy of $d^n \rightarrow d^{n+1}\underline{L}$, where $L$ denotes a hole in the anion $p$ valence band. For $U < \Delta$, the first ionization state is $d^{n-1}$, and for $U > \Delta$, it is $d^n\underline{L}$, assuming a $d^n$ ground state. For $Li_2TiO_3$, the $U$ value can be estimated as zero since $Ti^{4+}$ has no $3d$ electrons; Ti oxides normally have very large $\Delta$ values due to the large electronegativity difference between

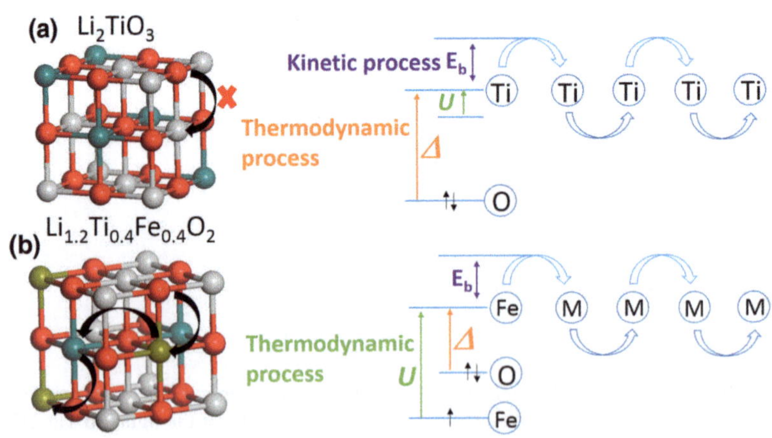


**Fig. 6.12** **a** and **b** show the charge-transfer processes for $Li_2TiO_3$ and $Li_{1.2}Ti_{0.4}Fe_{0.4}O_2$, respectively, during the redox process

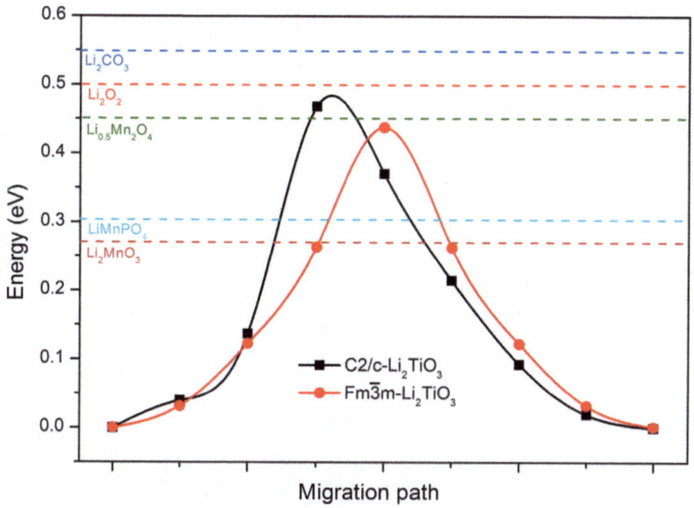


**Fig. 6.13** Calculated small hole polaron migration barriers for C2/c and Fm$\bar{3}$m $Li_2TiO_3$, with references to other electrodes for comparison

Ti and O. This leads to $U \ll \Delta$ in $Li_2TiO_3$ electrodes (Fig. 6.13a), regardless of the space group, such that compensating holes must be generated on $Ti^{4+}$, which can never supply *d* electrons, and thus leads to the thermodynamic inactivity. Because $U > \Delta$ for $Fe^{3+}$ (actually, $\Delta$ is zero since the band gap is zero) [36], holes will be generated on anion *p* states upon Fe substitution, and therefore, electron transfer can be thermodynamically activated via the O 2*p*-to-Fe 3*d* path (Fig. 6.13b). For a more special $Li_2MnO_3$ electrode, in which $Mn^{4+}$ lacks an empty 3*d* band but has a large band gap of 1.8 eV, oxygen redox activity is triggered due to the large *U* value of Mn and relatively smaller $\Delta$. This perfectly explains why $Li_2MnO_3$ exhibits considerable electrochemical redox activity, whereas other materials, such as $Li_2TiO_3$, do not.

To further evaluate this thermodynamic trigger in the charge-transfer process, we extended our study from oxides to sulfides (Fig. 6.14). Layered $Li_2TiS_3$ with C2/m symmetry was newly synthesized (Fig. 6.14a). This material shows little electrochemical activity (Fig. 6.14b, inset figure) despite a calculated band gap of only 1.2 eV, which would imply relatively good electron conductivity. However, after Co ions are incorporated into $Li_2TiS_3$, the resultant layered $Li_{1.2}Ti_{0.6}Co_{0.2}S_2$ exhibits a reversible capacity of ~200 mAh $g^{-1}$ (Fig. 6.14a, b). The redox process in $Li_{1.2}Ti_{0.6}Co_{0.2}S_2$ comprises mainly sulfur redox chemistry, as proven by the S *K*-edge XAS (Fig. 6.14c) and Ti and Co *K*-edge XAFS spectra (Fig. 6.15), as well as the charge-density difference map obtained by the DFT method (Fig. 6.14d). Coincidentally, the sulfur redox behavior in $Li_{1.2}Ti_{0.6}Co_{0.2}S_2$ presents selectivity similar to that of the oxygen redox behavior in $Li_{1.2}Ti_{0.4}Fe_{0.4}O_2$: The sulfur ions bonded only to Co ions are inclined to contribute considerable electrons (Fig. 6.14e), whereas, for sulfur surrounded by Ti ions, only a trace amount of charge outflow is observed (Fig. 6.14f).

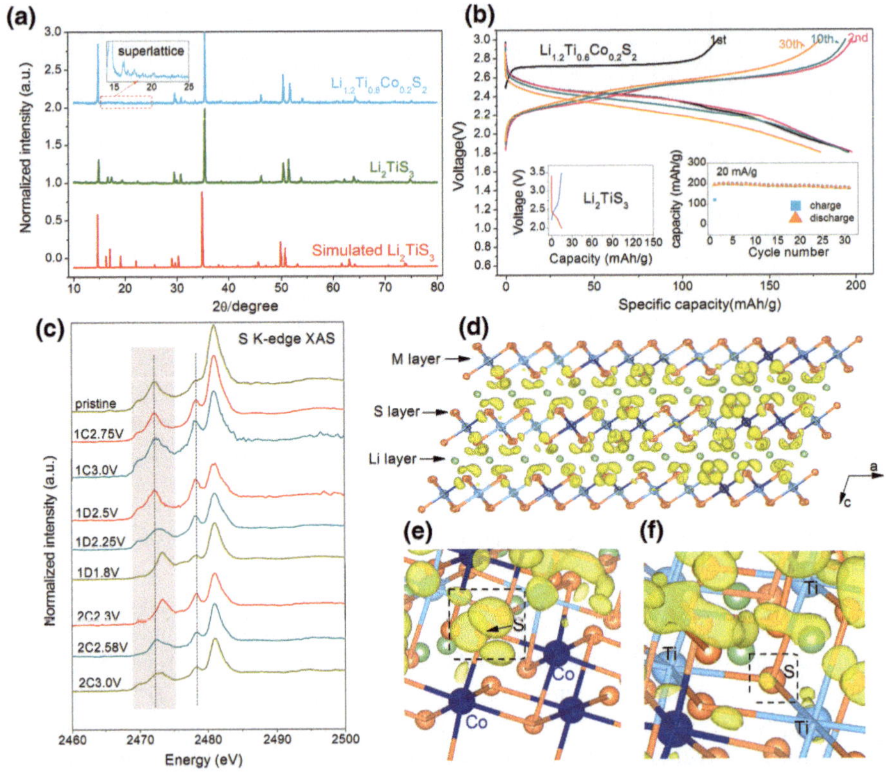


**Fig. 6.14** **a** XRD patterns of C2/m $Li_2TiS_3$ and $Li_{1.2}Ti_{0.6}Co_{0.2}S_2$, as well as the simulated C2/m $Li_2TiS_3$. **b** Electrochemical performance of C2/m $Li_2TiS_3$ and $Li_{1.2}Ti_{0.6}Co_{0.2}S_2$ at a current density of 20 mA $g^{-1}$. **c** S *K*-edge XAS spectra of $Li_{1.2}Ti_{0.6}Co_{0.2}S_2$ during the first cycle and second discharge. **d** Charge-density difference between $Li_{1.2}Ti_{0.6}Co_{0.2}S_2$ and $Ti_{0.6}Co_{0.2}S_2$. **e** and **f** show the selective redox behaviors of sulfur anions surrounded by different TM ions

These results strongly resemble the behaviors of $Li_2TiO_3$ and $Li_{1.2}Ti_{0.4}Fe_{0.4}O_2$ in which the thermodynamically favoured electron-transfer process of anionic redox is further proven: Co, with lower $\Delta$ than $U$ value, acts as a catalyst to lower the thermodynamic barrier and bridge the electron transfer in $Li_2TiS_3$. Another case that should be mentioned is Fm$\bar{3}$m-$Li_2TiS_3$, reported by Sakuda et al. [32], which was peculiar because it exhibited high electrochemical activity. The published results become explainable after we calculated that the band gap of Fm$\bar{3}$m-$Li_2TiS_3$ is zero (Fig. 6.16), indicating a zero $\Delta$ that permits the electron transfer of the sulfur redox process.

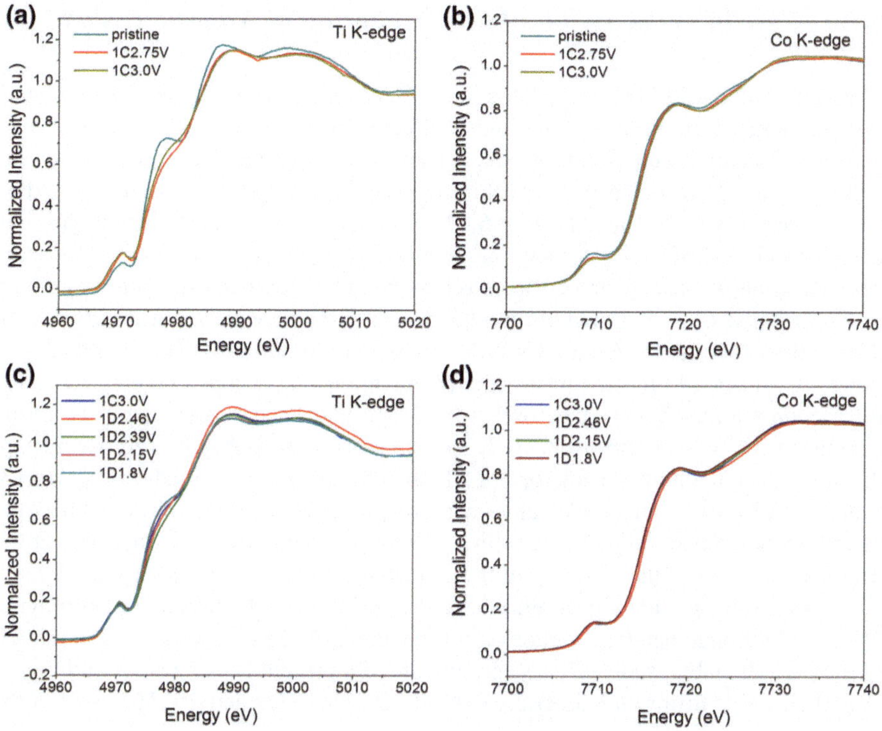


**Fig. 6.15** Ti and Co *K*-edge XAFS spectra for the first cycle of $Li_{1.2}Ti_{0.6}Co_{0.2}S_2$

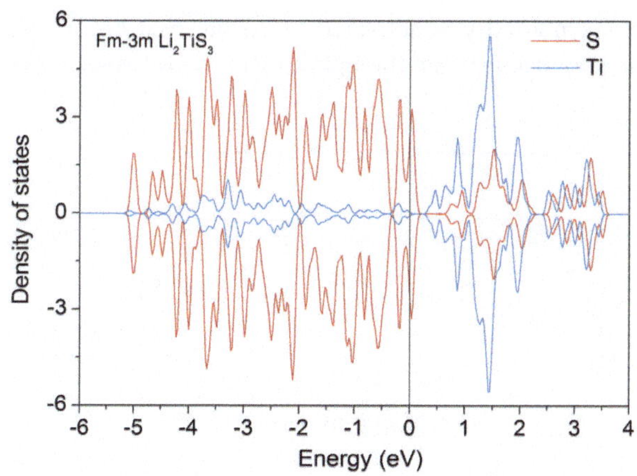


**Fig. 6.16** Calculated PDOS of $Fm\overline{3}m$ $Li_2TiS_3$

### 6.3.5 Further Discussion of the Triggering of Anionic Redox

The Mott–Hubbard $U$ value is related to the electronic configuration of the 3$d$ band, and the charge-transfer energy $\Delta$ is highly affected by the electronegativity difference between the cation and anion. Hence, for an overall charge/discharge process, the relative values of $U$ and $\Delta$ vary as lithium ions are extracted. This is exemplified by the charge process for $LiCoO_2$ (Fig. 6.17). Initially, $\Delta$ (4 eV) is slightly larger than $U$ (3.5 eV) in $LiCoO_2$ [37], which may induce a Co redox process at first. However, upon delithiation, $\Delta$ will largely decrease due to the increase in the electronegativity of Co, because $Co^{3+}$ is gradually oxidized. The $U$ value exceeds $\Delta$ at around $x$ = 0.06, with a changeover from a Co redox to an O redox process. This explains why previous studies all reported little valence change for $Co^{3+}$ ions [18, 19]. Actually, for a comparative $U$ and $\Delta$ before $x$ = 0.1, Co and O may participate the redox competitively. When $x$ approaches 0.3, $\Delta$ decreases to zero, implying a zero band gap; this is consistent with the insulator–metal transition previously found for $Li_{1-x}CoO_2$ near $x = 0.25$ [38]. For almost all anionic redox processes, the electrodes will become metallic when the anion $p$ DOS continuously passes the Fermi level, because of the decrease in $\Delta$ due to the levelling of the electronegativities of the metals and ligand.

A general understanding of anionic redox activation is facilitated by employing $U$ and $\Delta$, although accurate values of either can hardly be obtained. TMs to the left of the periodic table, such as Ti, V, Zr, Nb, and Ta, present low $U$ values and higher $\Delta$ in their oxide forms and generally exhibit cationic redox activity. However, if the cationic redox process is not available, the electrodes are inactive, as observed for $Li_2TiO_3$ and $Li_3NbO_4$. For the oxides of heavy TMs such as Mn, Fe, Co, Ni, Mo, Ru, and Ir, larger $U$ values (especially for high valence) than $\Delta$ are observed, and thus, anionic redox normally is activated when the oxygen 2$p$ states move to the Fermi level; this may occur when the upper metal $d$ electrons are depleted, such as

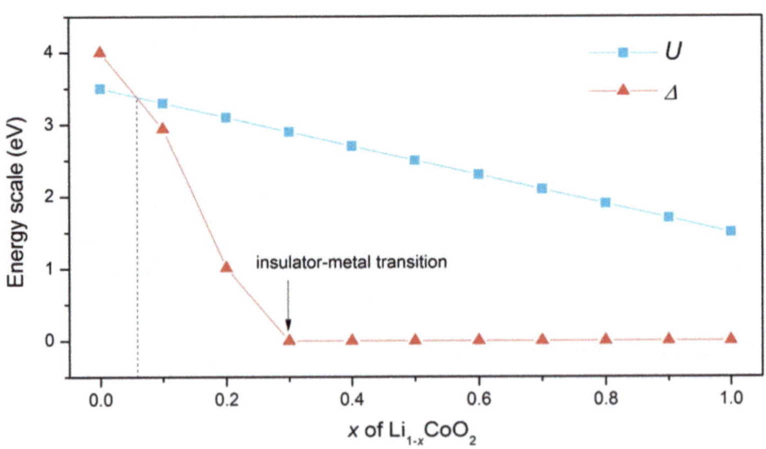


**Fig. 6.17** Dynamic evolution of $U$ and $\Delta$ values as a function of $x$ in $Li_{1-x}CoO_2$

in $Li_2RuO_3$. This rule is well confirmed by the activation of oxygen redox processes in M-doped $Li_3NbO_4$ (M = Mn, Fe, Co, and Ni) and the inactivation in V-doped $Li_3NbO_4$, as reported previously by Yabuuchi et al. [7, 8], who, more importantly, also fabricated a cation-disordered $Li_{1.2}Ti_{0.4}Mn_{0.4}O_2$ electrode that exhibited high oxygen redox activity [39]. The progressive oxygen redox activities in $Li_2TiO_3$, $Li_2MnO_3$, and $Li_2RuO_3$ can be typically understood via this theory. $Ti^{4+}$ in $Li_2TiO_3$ has a zero U value, far smaller than its large Δ value, which well explains its thermodynamic inactive, while for $Li_2MnO_3$, $Mn^{4+}$ possesses a slightly larger $U$ value (4–5 eV) than $\Delta$ value (3.3 eV), implying a restraint charge transfer from O to Mn. This is well verified by previous finds that $Li_2MnO_3$ only exhibits a little activity without special treatments, while being largely activated via methods like nanosizing, high temperature, and small current treatments. However, for $Li_2RuO_3$, a larger $U$ value than $\Delta$ value can be estimated, which results in greater oxygen redox activity than $Li_2TiO_3$ and $Li_2MnO_3$ under normal conditions. The evolution of Ti to Mn to Ru represents the light TM ions to heavy TM ions with larger and larger $U$ values relative to $\Delta$, and thus progressive charge-transfer ability in anionic redox reaction.

On the other hand, anions with lower electronegativities, such as sulfur and selenium, will also afford lower $\Delta$. This will trigger anionic redox activity in light TM chalcogenides, as exemplified by the cases of disordered $Li_2TiS_3$ and $Li_3NbS_4$, which exhibit considerable electrochemical activities. To some extent, due to their larger $U$ values, heavy TM ions can be applied as internal catalysts by doping to lower the thermodynamic barrier for anionic redox processes, as shown by Co-doped $Li_2O$ in a lithium-peroxide battery [40]. More amazingly, external catalysts comprising heavy TM ions, such as nano-$Co_3O_4$, are also available to stimulate oxygen redox activity in $Li_2O$, as recently reported [41].

## 6.4 Summary

Overall, our study shows that the substitution of $Fe^{3+}$ in inactive cubic $Li_2TiO_3$ triggers oxygen redox activity. The theoretical investigation demonstrated that electron transfer mainly occurs at oxygen atoms surrounded by Fe rather than Ti ions. This was perfectly explained by a thermodynamic activation of the charge-transfer process facilitated by Fe ions, which was discerned by the relative values of $U$ and $\Delta$. The extension to sulfides expanded the universality of this theory in providing a further general understanding of the anionic redox process. This finding will enhance efficiency when utilizing an anionic redox process to acquire higher capacity by composition design, such as by the combination of specific metal ions. It also introduces a new vision for exploiting a vast range of inactive electrodes such as $Li_2TiO_3$, $Li_2ZrO_3$, $Li_2SnO_3$, $Li_3TaO_4$, or even $LiAlO_2$ and $LiGeO_2$, by activating their oxygen redox activities via special metal substitution as catalysts. In a broader sense, this work will also benefit the development of lithium–oxygen and lithium–sulfur systems, which exclusively involve the charge-transfer processes of anions.

## References

1. Choi NS, Chen Z, Freunberger SA, et al. Challenges facing lithium batteries and electrical double-layer capacitors. Angew Chem Int Ed. 2012;51(40):9994–10024.
2. Chen J. Recent progress in advanced materials for lithium ion batteries. Materials. 2013;6(1):156–83.
3. Whittingham MS. Ultimate limits to intercalation reactions for lithium batteries. Chem Rev. 2014;114(23):11414–43.
4. Tarascon JM. Is lithium the new gold? Nat Chem. 2010;2(6):510.
5. Goodenough JB, Kim Y. Challenges for rechargeable Li batteries. Chem Mater. 2010;22(3):587–603.
6. Li H, Wang Z, Chen L, et al. Research on advanced materials for Li-ion batteries. Adv Mater. 2009;21(45):4593–607.
7. Yabuuchi N, Takeuchi M, Nakayama M, et al. High-capacity electrode materials for rechargeable lithium batteries: $Li_3NbO_4$-based system with cation-disordered rocksalt structure. Proc Natl Acad Sci. 2015;112(25):7650–5.
8. Yabuuchi N, Takeuchi M, Komaba S, et al. Synthesis and electrochemical properties of $Li_{1.3}Nb_{0.3}V_{0.4}O_2$ as a positive electrode material for rechargeable lithium batteries. Chem Commun. 2016;52(10):2051–4.
9. Sathiya M, Rousse G, Ramesha K, et al. Reversible anionic redox chemistry in high-capacity layered-oxide electrodes. Nat Mater. 2013;12(9):827–35.
10. Sathiya M, Ramesha K, Rousse G, et al. High performance $Li_2Ru_{1-y}Mn_yO_3$ ($0.2 \leq y \leq 0.8$) cathode materials for rechargeable lithium-ion batteries: their understanding. Chem Mater. 2013;25(7):1121–31.
11. Sathiya M, Abakumov AM, Foix D, et al. Origin of voltage decay in high-capacity layered oxide electrodes. Nat Mater. 2015;14(2):230–8.
12. Ma J, Gao Y, Wang Z, et al. Structural and electrochemical stability of Li-rich layer structured $Li_2MoO_3$ in air. J Power Sources. 2014;258:314–20.
13. Ma J, Zhou Y-N, Gao Y, et al. Feasibility of using $Li_2MoO_3$ in constructing Li-rich high energy density cathode materials. Chem Mater. 2014;26(10):3256–62.
14. Lee J, Urban A, Li X, et al. Unlocking the potential of cation-disordered oxides for rechargeable lithium batteries. Science. 2014;343(6170):519–22.
15. Lee J, Seo D-H, Balasubramanian M, et al. A new class of high capacity cation-disordered oxides for rechargeable lithium batteries: Li–Ni–Ti–Mo oxides. Energy Environ Sci. 2015;8(11):3255–65.
16. Wang R, Li X, Liu L, et al. A disordered rock-salt Li-excess cathode material with high capacity and substantial oxygen redox activity: $Li_{1.25}Nb_{0.25}Mn_{0.5}O_2$. Electrochem Commun. 2015;60:70–3.
17. Van der Ven A, Aydinol MK, Ceder G, et al. First-principles investigation of phase stability in $Li_xCoO_2$. Phys Rev B. 1998;58(6):2975–87.
18. Montoro LA, Abbate M, Rosolen JM. Changes in the electronic structure of chemically deintercalated $LiCoO_2$. Electrochem Solid-State Lett. 2000;3(9):410–2.
19. Mesilov VV, Galakhov VR, Gizhevskii BA, et al. Charge states of cobalt ions in nanostructured lithium cobaltite: X-ray absorption and photoelectron spectra. Phys Solid State. 2013;55(5):943–8.
20. Yukinori K, Yang-Soo K, IsaoTanaka, et al. Changes in chemical bondings by Li deintercalation in $LiMO_2$ (M=Cr, V, Co and Ni). Jpn J Appl Phys. 1999;38(4R):2024.
21. Graetz J, Hightower A, Ahn CC, et al. Electronic structure of chemically-delithiated $LiCoO_2$ studied by electron energy-loss spectrometry. J Phys Chem B. 2002;106(6):1286–9.
22. Wolverton C, Zunger A. First-principles prediction of vacancy order-disorder and intercalation battery voltages in $Li_xCoO_2$. Phys Rev Lett. 1998;81(3):606–9.
23. Yoon WS, Kim KB, Kim MG, et al. Oxygen contribution on Li-ion intercalation−deintercalation in $LiCoO_2$ investigated by O K-edge and Co L-edge X-ray absorption spectroscopy. J Phys Chem B. 2002;106(10):2526–32.

24. Mizokawa T, Wakisaka Y, Sudayama T, et al. Role of oxygen holes in $Li_xCoO_2$ revealed by soft X-ray spectroscopy. Phys Rev Lett. 2013;111(5):056404.
25. Wang Z, Wang Z, Peng W, et al. Structure and electrochemical performance of $LiCoO_2$ cathode material in different voltage ranges. Ionics. 2014;20(11):1525–34.
26. Chen C-H, Hwang B-J, Chen C-Y, et al. Soft X-ray absorption spectroscopy studies on the chemically delithiated commercial $LiCoO_2$ cathode material. J Power Sources. 2007;174(2):938–43.
27. Dahéron L, Dedryvère R, Martinez H, et al. Electron transfer mechanisms upon lithium deintercalation from $LiCoO_2$ to $CoO_2$ investigated by XPS. Chem Mater. 2008;20(2):583–90.
28. Aydinol MK, Kohan AF, Ceder G, et al. Ab initio study of lithium intercalation in metal oxides and metal dichalcogenides. Phys Rev B. 1997;56(3):1354–65.
29. Shang SL, Wang Y, Mei ZG, et al. Lattice dynamics, thermodynamics, and bonding strength of lithium-ion battery materials LiMPO4 (M = Mn, Fe Co, and Ni): a comparative first-principles study. J Mater Chem. 2012;22(3):1142–9.
30. Xiao R, Li H, Chen L. Density functional investigation on $Li_2MnO_3$. Chem Mater. 2012;24(21):4242–51.
31. Tarascon JM, Armand M. Issues and challenges facing rechargeable lithium batteries. Nature. 2001;414(6861):359–67.
32. Sakuda A, Takeuchi T, Okamura K, et al. Rock-salt-type lithium metal sulphides as novel positive-electrode materials. Sci Rep. 2014;4:4883.
33. Saubanère M, McCalla E, Tarascon JM, et al. The intriguing question of anionic redox in high-energy density cathodes for Li-ion batteries. Energy Environ Sci. 2016;9(3):984–91.
34. Seo DH, Lee J, Urban A, et al. The structural and chemical origin of the oxygen redox activity in layered and cation-disordered Li-excess cathode materials. Nat Chem. 2016;8(7):692–7.
35. Sawatzky GA, Allen JW. Magnitude and origin of the band gap in NiO. Phys Rev Lett. 1984;53(24):2339–42.
36. Zaanen J, Sawatzky GA, Allen JW. Band gaps and electronic structure of transition-metal compounds. Phys Rev Lett. 1985;55(4):418–21.
37. van Elp J, Wieland JL, Eskes H, Kuiper P, Sawatzky GA, de Groot FMF, Turner TS. Electronic structure of CoO, Li-doped CoO, and $LiCoO_2$. Phys Rev B. 1991;44:6090.
38. Menetrier M, Saadoune I, Levasseur S, et al. The insulator-metal transition upon lithium deintercalation from $LiCoO_2$: electronic properties and $^7Li$ nmr study. J Mater Chem. 1999;9(5):1135–40.
39. Yabuuchi N, Nakayama M, Takeuchi M, Komaba S, Hashimoto Y, Mukai T, Shiiba H, Sato K, Kobayashi Y, Nakao A, Yonemura M, Yamanaka K, Mitsuhara K, Ohta T. Origin of stabilization and destabilization in solid-state redox reaction of oxide ions for lithium-ion batteries. Nat Commun. 2016;7:13814.
40. Ogasawara Y, Hibino M, Kobayashi H, Kudo T, Asakura D, Nanba Y, Hosono E, Nagamura N, Kitada Y, Honma I, Oshima M, Okuoka S-I, Ono H, Yonehara K, Sumida Y, Mizuno N. Charge/discharge mechanism of a new Co-doped $Li_2O$ cathode material for a rechargeable sealed lithium-peroxide battery analyzed by X-ray absorption spectroscopy. J Power Sources. 2015;287:220.
41. Zhu Z, Kushima A, Yin Z, Qi L, Amine K, Lu J, Li J. Anion-redox nanolithia cathodes for Li-ion batteries. Nat Energy. 2016;1:16111.

# Chapter 7
# Conclusions and Outlooks

## 7.1 Conclusions

Anionic redox process is the crucial point in Li-rich rock-salt type oxide cathode materials and plays important role in designing new high-capacity electrodes. Deeply investigating the anionic redox mechanism will substantiate the scientific basis of high-capacity cathode materials. The main contents and conclusions of this thesis include:

1. We have fabricated a kind of polyanion-modified lithium-rich manganese-based layered oxide, by introducing boracic polyanions into the layered structure to tune its electronic structure. The as-prepared electrode exhibits excellent cycling performance, specifically, showing 300 $mAh{\cdot}g^{-1}$ when cycled to 80th cycle under 0.1C rate current and 89% retention cycled to 300th cycle under 0.3C rate, with also higher thermal stability and charge/discharge potentials. Combined series of techniques, we have shown that the improvement should be ascribed to the incorporation of boracic polyanions, which lowers the overlap of M 3d (M = transition metal) and O 2p states, thus reducing the oxygen oxidation and resulting a more stable structure.
2. Based on the model of $Li_2RuO_3$, the geometric and electronic structural evolutions of the lithium-rich high-capacity electrode were systematically investigated, with the aid of in situ XRD, in situ XAFS, and in situ TEM & EELS techniques. Further facilitated by DFT calculations, we found that the excessive lithium ions residing in the transition metal layer actually enhances the robustness and flexibility of the local structure, which enables the O–O bonds to contract forming peroxide species and the local structure to distort to relieve the energy of the whole system, thus stabilizing the local structure during oxygen redox process.
3. We have successfully proved that the reversibility of the oxygen redox in $Li_2RuO_3$ can be tuned via doping certain amount of boron atoms into the layered structure, which varies the electronic structure of Ru–O system. The results show that after the doping of boron atoms, $Li_2RuO_3$ exhibits more capacity contributed by



B. Li, *Studies on Anionic Redox in Li-Rich Cathode Materials of Li-Ion Batteries*, Springer Theses, https://doi.org/10.1007/978-981-13-2847-3_7

Ru redox, while the capacity corresponding to oxygen redox is reduced. As a result, the modified $Li_2RuO_3$ presents great cycling stability with more reversible oxygen redox. Specifically, $Li_2RuB_{0.15}O_3$ can deliver 219 $mAh \cdot g^{-1}$ in the first cycle and can be cycled to 100th cycle with the retention of 86.7%.

4. The substitution of $Fe^{3+}$ in cubic $Li_2TiO_3$ will trigger the activity of oxygen redox, though $Fe^{3+}$ did not participate in the redox process. We have successfully made clear the activation mechanism by introducing the $U$-$\Delta$ theory derived from Zaanen–Sawatzky–Allen model and expanding it to a general level in the range of both lithium-ion and lithium–oxygen systems. Though high-covalent theory can explain the origin of anionic redox in certain extent, our theory shows more superiority in the manner of specific, theoretical, general, and quantitative in uncovering the activation mechanism.

## 7.2 Outlooks

It should be noted that the anionic redox can provide multiple electron processes and thus lead to high capacity, but on the other hand it will also destabilize the structure and cause severe capacity loss or short cycle life. Therefore, the further development of high-capacity electrodes based on anionic redox, as exemplified by the lithium-rich systems, should be implemented under the circumstance of controlling the anionic redox to be a very reversible level, as well as the host structure. The possibly induced voltage hysteresis by anionic redox is also an unresolved problem in obtaining applicable electrodes. At last, the characterization of the anionic redox should play a more vital role in the future research, and the combination work of detecting skills is more preferred due to their respective advantages.

The redox mechanism of rechargeable lithium batteries is started from conventional sole metal center in conventional $LiMO_2$ oxides. With the increase of lithium ions on the expense of metal content, the redox involves anions, such as the case of $Li_2MO_3$. When metal ions are electrochemically inactive, sole oxygen redox may be excited, as observed in Y-doped $Li_8ZrO_6$. Based on this trend, if lithium substitute all of the metal in the oxides, an extreme case of the redox is $(O^{2-})_2/O_2$ in the $Li_2O_2$ electrode of lithium–oxygen batteries. Note that the high overpotential and the activation voltage of the oxygen redox in lithium–oxygen system are very similar with that in lithium-ion batteries. In view of this, we believe that the ultimate station of the exploration of anionic redox in lithium-ion batteries may approach or reach the lithium–oxygen batteries. More oxygen redox contribution, higher the capacity, and less stability of the structure. The future strategy in designing electrodes may consider compromising these two battery systems, that tuning the relative content of lithium and metal in chalcogenides, thus controlling the contribution of anionic redox. Since catalysts are required in facilitating the lithium–oxygen reactions, proper active metal with catalytic effect can be doped into some electrochemical inactive compounds, such as layered $Li_2TiO_3$ and $Li_3NbO_4$, to activate their oxygen redox activity.

MIX
Papier aus verantwortungsvollen Quellen
Paper from responsible sources
FSC® C105338

If you have any concerns about our products,
you can contact us on
**ProductSafety@springernature.com**

In case Publisher is established outside the EU,
the EU authorized representative is:
**Springer Nature Customer Service Center GmbH**
**Europaplatz 3, 69115 Heidelberg, Germany**

Printed by Libri Plureos GmbH
in Hamburg, Germany